ANYONE CAN LEARN MATH

ANYONE CAN LEARN MATH

Helping High Risk Learners

Aileen Midori Yamate, Ed. D.

iUniverse, Inc.
New York Lincoln Shanghai

Anyone Can Learn Math
Helping High Risk Learners

Copyright © 2006 by Aileen Yamate

iUniverse books may be ordered through booksellers or by contacting:

iUniverse
2021 Pine Lake Road, Suite 100
Lincoln, NE 68512
www.iuniverse.com
1-800-Authors (1-800-288-4677)

ISBN-13: 978-0-595-27736-0 (pbk)
ISBN-13: 978-0-595-65710-0 (cloth)
ISBN-10: 0-595-27736-5 (pbk)
ISBN-10: 0-595-65710-9 (cloth)

Printed in the United States of America

DEDICATION

Sandra Hollingsworth, Ph.D.
Graduate School of Education
University of California, Berkeley

CONTENTS

INTRODUCTION

For 25 years I was a classroom teacher, a mentor teacher, and a supervisor of student teachers. For most of those years, I taught in two of the most challenging school environments—low-income and poverty areas in the Bay Area of Northern California (Richmond) and Southern California (South Central Los Angeles ["Watts"]).

I have found a way to reach today's students, especially those students who are academically and socially challenging to teach—our "at risk" youths. The students that I taught were labeled as "unmotivated," "challenging youths with an attitude,"—"I dare you to teach me".

Despite insurmountable odds, most of my students came through for me. They transformed themselves from unmotivated, rebellious, and problem students to achievers, scoring above average on their SAT scores. Some of them have gone on to college on scholarships; some have steady jobs in the banking system and retail stores.

This book is sequentially designed; therefore, the teacher can quickly determine where to focus instruction and master. It requires constant analysis of deficiencies and weaknesses found in the students' learning, and providing feedback on the spot. Following this protocol is crucial, especially if you teach "at risk" learners.

AUTHOR'S NOTE

Welcome to the Yamate Method of teaching mathematics!

Because my method is "learner centered "rather than "teacher centered," what I offer is not "grade-level-centered" or strictly rote learning and memorization. It is based upon (a) the readiness of the learners as they demonstrate that they have mastered each essential step in a mathematical computation, (b)assures that students know for themselves that they have mastered a particular computation, (c) the repetition in learning mathematics, (d) opportunities for your students to tell you how they are learning, (e) requires your willingness to listen and be the student so-to-speak, (f) constantly analyze and provide feedback to your students, (g) be as creative as you can, especially when you are teaching, (h) demanding situation, but the rewards are great, seeing confidence and self-esteem develop in your students, (i) building credibility, trust, and loyalty.

The Yamate Strategy

This book provides

1. computational thinking, reasoning, and solving complex problem

2. basic calculations anyone can perform: addition, subtraction, multiplication, and division with whole numbers, decimal numbers, and fractions

3. set of objectives, explanations, step-by-step examples, and practice exercises

4. sixty-five basic errors that are common in basic mathematics that you can focus on and master

5. analysis of errors, diagnosis, and techniques for teaching

6. presents list of analysis of errors in each chapter.

PLAN OF THE BOOK

"Anyone Can Learn Math" is intended to be the first in a series, beginning with arithmetic and progressing to higher forms of mathematics. It is designed to support teachers who teach arithmetic to "at-risk" students—students who have a long history of failure, having been passed on from grade to grade without mastering the essentials to prepare them for tackling higher levels of mathematics.

"At-risk" students are those students who show up in class unmotivated with an "I dare you to teach me" attitude. I identify techniques I used in teaching my students a particular math concept—some of them are out of a textbook; some of them are created on the spot. I describe my interaction with my students as I take them through a particular step in solving a problem, which demonstrates that concept.

Mathematics is a vehicle (as numerical literacy) to practice thinking, reasoning, and judging, which are important to function in life. Mathematics is supposed to be a unifier in that it is the only subject taught whose principles are universal and universally taught (financial, statistical, computational, etc.). As it is said in professional circles, mathematics may be the only non-culturally determined behavior that is taught to all children in the school system. It is a constant. This is especially important since we have entered the 21st century's global society.

---

[1] "At risk" refers to any population of students who are experiencing sufficient difficulties in school that they are likely to become school failures. The educational reform movement of the 1980s termed the education of ethnic groups as "culturally deprived or disadvantaged learners" and the label "at risk" was popularized as a category for State and Federal Educational Funding (National Commission on Excellence in Education).

CHAPTER 1

Set-Up 1: ADDITION OF WHOLE NUMBERS

This objective requires students to solve complex column addition requiring the composition of lower value unit to higher value unit—regrouping.

Problem Sample Set

a)
$$
\begin{array}{r}
2,6\,9\,1 \\
+\ 8,3\,2\,9 \\
\hline
\end{array}
$$

b)
$$
\begin{array}{r}
5\,8,3\,4\,7 \\
+\ \ 4,2\,1\,5 \\
\hline
\end{array}
$$

c)
$$
\begin{array}{r}
3,9\,2\,6 \\
2,5\,1\,8 \\
6,0\,2\,9 \\
+\ \ \ 7,6\,8\,1 \\
\hline
\end{array}
$$

d) $1,1\,8\,4 + 6,2\,9\,8 + 3,5\,6\,8$

e) $2\,3,0\,0\,4 +\ 1\,2,0\,4\,6 + 5\,1,2\,3\,1$

Analysis of Errors

1. Computation error.

2. Carrying errors (carried wrong number).

3. Weakness in combinations.

4. Counting.

5. Vocalizing work.

6.Losing place in column addition.

7. Losing place in horizontal addition.

8. Have no regular procedure in working addition—proceeds one column from the top and the next from the bottom, forgetting to add the carried number.

9. Split the numbers to get easier combination.

Assessing Students' Mastery of Basic Number Facts in Addition (while carrying on dialogue with students).

"Today, I am going to pull the strings tight on you by giving you a test on basic facts in addition. I will time you and when I say stop, you will mark a line across your paper. Then, you will finish the rest of your paper. You may sit near your friend, if you wish."

(A spontaneous, correct response is sought from every student.)

"Oh no, I'm not going to sit near you. Go back, Kevin. I'm going to get an 'A' on this test," said Charles.

"Jimmy, move to the back of the room. I can't concentrate when you mumble," said Jennifer.

(Only 5 out of the 35 students sat together. It seems that they felt the need to concentrate and do well on the test. I am a new teacher, and they are looking at me with renewed hope.)

"We are going to practice this every day for five minutes until we get 100 percent correct. You will each keep a record of your scores in your folders. At the end of the week, you will make a *graph* of your progress. Maybe we will be through with this testing before the end of the week."

"Yeah. How many of you are going to get an 'A'?" asked Jamal.

(All hands went up. Testing started at a level they are confident and sure to know.)

First Day of Assessment.

1. Drill in basic facts was given both orally and in written form (adding three to five whole numbers in a column). It was orally paced so that finger counting was not possible.

2. Papers were corrected immediately after the test, and each student corrected his or her paper. All error combinations were noted on the back of his or her paper to be studied.

(The drill was paced to meet the needs of the various students—additional time given to slower workers to complete the test. Speed will come with practice.)

Place Value and Renaming (Carrying/regrouping to higher value).

"Students, please copy the place value chart on the chalkboard onto a piece of paper and title it *'Notes.'* I would like you to keep notes in this class when I lecture so you can have a reference sheet. Anything that I write on the chalkboard that you want to remember—a procedure or an algorithm, please copy it down. Besides, I may give you a test and say, 'All right, you may use your notes on today's test.' This paper will remain in your folders."

(Students need to develop a habit of note taking. Some students have displayed difficulty in visual or auditory perception, in organizing or interpreting information, or in sorting or retrieving presented information.)

The following problems were written on the chalkboard to solve as a class:

```
       2
    1 2 6 8        8 5 2   (Addend)
    1 1 9 5        5 9 5   (Addend)
    2 0 3 9        4 6 0   (Addend)
    3 2 4 7        3 3 5   (Addend)
  + 9 7 8 0      + 1 7 4   (Addend)
        9                  (Sum)
```

(Three different answers were obtained from the class on the first problem and two different answers on the second.)

"Since we have different answers, why don't we work the problems together? Please respond. Eight plus five."

Students: "Thirteen."

"Plus nine."

Students: "Twenty-two."

"Plus seven."

Students: "Twenty-nine."

"Plus zero."

Students: "Twenty-nine."

"Say TWENTY-nine, class. How many tens in twenty-nine?"

Students: "Two tens."

"How many ones?"

Students: "Nine ones."

"Where do you place the two?"

Students: "At the top of the next column."

"What is the value of that column?"

Students: "Tens."

"Put a circle around the 2 as a reminder. Now, write the 9 as the sum in the ones column."

Kenya: "Oh, that's what I did. I put the 2 down first and carried the 9. I always get mixed up. I think that I carried the right number, but I seem to put down the wrong number. If I can 'think' out loud like we just did, I don't think I'd make this kind of mistake."

(A teacher may tell a student to be quiet when doing his or her work; however, a student may need to think aloud, to hear himself or herself while writing.)

Teacher: "The first row of numbers look fine, Michael, but what happened to the other rows. The numbers seem to be dancing in between the upper row of numbers. The digits need to be aligned in each row. Let's use a *graph* paper and write these numbers down in the correct column and row. It looks as though your error came from not having your numbers in clear columns."

(Michael's writing is illegible.)

Derrick: "There are too many numbers to add, Mrs. Yamate. I got mixed up keeping track."

"Here are two other ways of adding a column without carrying that you might try."

Adding from right to left column OR Adding from left to right column

```
    1 2 6 8                  8 5 2
    1 1 9 5                  5 9 5
    2 0 3 9  (addend)        4 6 0  (addend)
    3 2 4 7                  3 3 5
  + 9 7 8 0                + 1 7 4
        2 9              2 1
        3 0                  3 0    (partial sum)
    1 2      (partial sum)  +     1 6
  + 1 6                      2 4 1 6  (sum)
    1 7 5 2 9  (sum)
```

Michael: "Hey, this is easier. I don't have to worry about carrying, and it is fun!"

"How can we check our work, class?"

Samuel: "Do it over again."

Jamie: "Hey, you can break up the numbers in the same column. Just combine the numbers you know first, like 9 and 8 equals 17, and 7 and 5 equals 12; then add 17 and 12 to get the total sum 29. This way is easier and faster for me."

Samuel: "No, add from bottom to the top. Go backwards."

"Jamie, will you please come and show us your method on the board?"

Michael: "Hey, why does different grouping of numbers in a column work?"

Jamie: "That's just the way addition is. I've been doing this all the time. I found that breaking up the column or changing the order of adding the numbers does not change the sum."

"Class, please write this down in your notes:

Properties of Addition

1. Changing the order of two addends: $5 + 3$ or $3 + 5$ does not change the sum. This is called the commutative property or law of addition.

2. Changing the groupings or association of three numbers to be added does not change the sum. This is called the associative property or law of addition.

$$5 + 3 + 2 = (5 + 3) + 2 \qquad 5 + 3 + 2 = 5 + (3 + 2)$$
$$= \quad 8 \ + 2 \qquad\qquad\qquad = 5 + \quad 5$$
$$= \quad 1\,0 \qquad\qquad\qquad\qquad = \quad 1\,0$$

3. Zero (0) added to a whole number gives a sum that is the other addend: $6 + 0 = 6$. This is called the identity element for addition.

"All right, class. Please check your work now. Choose a method that is easiest for *you*."

The following examples were written on the chalkboard for further understanding of the process:

Shortcut method:

```
  I   I          8 + 4 = 12
  2  6  8        1 (ten carried) + 6 + 3 = 10
+ 2  3  4        1 (hundred carried) + 2 + 2 = 5
  5  0  2
```

Expanded notation: Place Value in number and words.

```
  2 6 8 = 2 0 0 + 6 0  + 8  = 2 hundreds + 6 tens + 8 ones
+ 2 3 4 = 2 0 0 + 3 0  + 4  = 2 hundreds + 3 tens + 4 ones
        = 4 0 0 + 9 0  + 12 = 4 hundreds + 9 tens + 12 ones
        = 4 0 0 + 1 0 0 + 2 = 4 hundreds + 10 tens + 2 ones
        = 5 0 0 +     2      = 5 hundreds +  0 tens + 2 ones
        = 5 0 2              = 5 hundred two
```

Expanded notation using powers:

$$2\ 6\ 8 = 2(10^2) + 6(10^1) + 8(10^0)$$
$$= 2(100) + 6(10) + 8(1)$$
$$= 200 + 60 + 8$$
$$= \text{two hundred sixty-eight (word equivalent)}$$

PLUS

$$2\ 3\ 4 = 2(10^2) + 3(10^1) + 4(10^0)$$
$$= 2(100) + 3(10) + 4(1)$$
$$= 200 + 30 + 4$$
$$= \underline{\text{two hundred thirty-four (word equivalent)}}$$
$$400 + 90 + 12 = 502$$

(The example shows the multiplication of each *digit* by the *value* of each *place*, a *power of ten* in our decimal system, and the addition of partial products to obtain the product 502.)

"All right class, I am going to teach you another alternative method of adding columns of numbers. This is called 'Arithmetic of Tens' [2]

---

[2] Fulkerson, Elbert. "Adding by Tens", Arithmetic Teacher, XX (March, 1963), pp 139-40.

```
    4 6 8
    7 9 5
    8 0 9
+   9 8 6
  3 0 5 8
```

Beginning at the top right in the example, 8 plus 5 equals 13, which can be renamed as 1 ten and 3 ones. A *line* is drawn through the 5 to show that it as the last digit used in obtaining ten. The ten does not need to be held in mind because the line represents it for us. Now, with the 3 that was left over, add until another ten is obtained. In this case 3 plus 9 equals 12, which can be renamed as 1 ten plus 2. *Another line* is drawn (through the 9) to indicate another ten. The sum of the 2 left over and 6 is 8, which is recorded as the ones digit at the bottom of the column.

The *two lines* which are drawn in the ones column represent *2 tens*, and addition begins in the tens column by adding these 2 tens to the 6 tens and continuing until there is a sum greater than 10 tens; 2 tens plus 6 tens plus 9 tens equals 17 tens.

A *line* is drawn through the 9 to represent *10 tens*, and 7 tens remain. Seven plus 8 equals 15; a *line* is drawn through the 8, and the 5 is recorded as the tens digit at the bottom.

Each line represents 10 tens or 1 hundred. There are *two* such *lines*, so *2 hundreds* are added to the 4 hundreds in the next column, and addition proceeds similarly."

"I'm not going to add like that. It's going to confuse me," said Marsha.

"I think this is a neat way to add, Mrs. Yamate. I'm going to practice this. Will you give us some problems to take home to practice?" asked Walter.

(Choice given according to student's need. Examples of practice worksheets passed out:

```
  1 2 6 8          2 3 6 9
  3 8 8 7          1 8 9 5
  6 5 4 6          2 0 3 9
+ 1 3 6 3        + 1 1 8 0
```

Skill Practice (Noting specific problem)

Column addition involving carrying numbers.

$$\begin{array}{r} 3\ 6\ 8 \\ +\ \quad 2 \\ \hline \end{array}$$ Carrying in adding by ending.

$$\begin{array}{r} 3\ 9\ 6 \\ +\ \quad 4 \\ \hline \end{array}$$

$$\begin{array}{r} 6\ 4\ 7 \\ +\ \quad 3 \\ \hline \end{array}$$

$$\begin{array}{r} 4 \\ 8\ 9 \\ +\ 6\ 6\ 2 \\ \hline \end{array}$$ Carrying to a vacant place.

$$\begin{array}{r} 2 \\ 2\ 6 \\ +\ 8\ 9\ 4 \\ \hline \end{array}$$

$$\begin{array}{r} 8\ 0\ 8 \\ 4\ 0\ 5 \\ +\ 1\ 0\ 1 \\ \hline \end{array}$$ Carrying into zero.

$$\begin{array}{r} 2\ 0\ 8 \\ 8\ 0\ 8 \\ +\ 2\ 0\ 9 \\ \hline \end{array}$$

$$\begin{array}{r} 1\ 9\ 1\ 8 \\ 8\ 1\ 5\ 6 \\ +\ 6\ 4\ 2\ 7 \\ \hline \end{array}$$ Carrying in alternate places.

$$\begin{array}{r} 7\ 2\ 4\ 5 \\ 2\ 5\ 2\ 6 \\ +\ 3\ 4\ 1\ 9 \\ \hline \end{array}$$

<u>Horizontal addition with carrying and regrouping to higher value</u>

6 + 3 7 5 + 1 5

4 2, 3 4 7 + 4, 2 1 5

5 7, 7 3 1 + 2, 5 7 8 + 4 3 2

"I can't add like that. I got lost in that math test because the problem was written sideways," said Jerald. "I didn't do them."

"Why didn't you rewrite the problem vertically in columns?"

"No one told me I could get scratch paper," said Jerald.

"You might keep track of the number in each place by putting a *check* mark over the digits as you add horizontally; for example, put a square around the ones digit, a circle around the tens digit, and a *triangle* around the hundreds digit. See what works for *you*."

"Please check your answers using a method that you are comfortable with." (Checking their work is important—for reinforcing correctness of their solution.)

"Let's study the Place Value Chart. Our base ten (decimal) system of numeration (number symbols, 0, 1, 2, 3, 4, 5, 6, 7, 8, 9), is grouped into sets of ten or multiples of ten—determined by its place (position) in any one place to make 1 in the next higher place. For example, it takes 10 ones to make 1 ten, 10 tens to make 1 hundred (each place being a power of ten, 10^0, 10^1, 10^2, 10^3, etc.). For example, the value of 0 in <u>30</u> is 0 ones, but the value of 3 in 30 is 3 times 10, or thirty ones."

Place Value Chart and Composition of Ten

Two Million Four Hundred Seventy-three Thousand Six Hundred Eighteen

millions	hundred-thousands	ten-thousands	thousands	hundreds	tens	ones
2	4	7	3	6	1	8

"The value of each digit in our base ten (decimal) system of numeration utilizes ten digits: 0, 1, 2, 3, 4, 5, 6, 7, 8, and 9. The value of a given digit is identified by its place (position)—for example, it takes 10 ones to make 1 ten, 10 tens to make 1 hundred (each place being a power of ten, 10^0, 10^1, 10^2,

10^3, etc.). For example, the value of 0 in <u>30</u> is 0 ones, but the value of 3 in 30 is 3 times 10, or thirty ones."

SUMMARY

LIST OF COMMON ERRORS:

1. Weakness in combination.

$$\begin{array}{r} 9 \\ +\ 8 \\ \hline 1\ 6 \end{array}$$

2. Carrying errors.　Carries wrong number.

$$\begin{array}{r} 1\ 9 \\ +\ 6 \\ \hline 6\ 1 \end{array}$$

Forgets to carry.

$$\begin{array}{r} 5\ 9 \\ +2\ 7 \\ \hline 7\ 6 \end{array}$$

Forgets to carry and add.

$$\begin{array}{r} 5\ 9 \\ +\ 2\ 7 \\ \hline 7\ 1\ 6 \end{array}$$

3. Same digit added to both columns.

$$\begin{array}{r} 2\ 4 \\ +\ 4 \\ \hline 6\ 8 \end{array}$$

4. Difficulty with zero.

$$\begin{array}{r} 3\ 0 \\ +9\ 4 \\ \hline 1\ 2\ 0 \end{array}$$

Mistakes $4 + 0$ with $4 \times 0 = 0$.

5. Losing place in column addition.

6. Losing place in horizontal addition.

7. Have no regular procedure in working addition--proceeds one column from the top and the next from the bottom, forgetting to add the carried number.

8. Splitting the numbers to get easier combination.

CHAPTER 2

Set-Up 2: SUBTRACTION OF WHOLE NUMBERS.

This objective requires students to solve problems involving decomposition of digits of higher value.

Problem Sample Set:

a)
$$532 \\ -\ 76$$

b)
$$6328 \\ -4794$$

c)
$$9080 \\ -6759$$

d)
$$5000 \\ -3373$$

e) $33,470 - 21,369$

Effective Strategy—<u>Mastery of The Basic Subtraction Facts</u> (while carrying on dialogue with students.)

"Pupils, wind up your brains and let's try taking a test on the basic facts of subtraction."

"What did you say, Mrs. Yamate? Pupils? What are you talking about?"

"Pupils of your eyes?" asked Stephen.

"No, pupils...students like you. Haven't you ever heard of that word?"

"It's news to us...*students* are *pupils*?" asked Stephen.

"Well, they used to call *students, pupils,* when I went to grammar school. They still do in some places. Okay, let's get started on this subtraction worksheets. I will time you and when I say stop, you will mark a line across your paper as you did in addition. Then, you will finish the rest of your paper. Anyone want to sit with a friend?"

(Students given a choice.)

"No. I'm going to get more As in this class. Man, my mom is really surprised I'm getting good grades in this class...that's because I've been studying like you told us to, Mrs. Yamate," said Steven. "We could get to them hard books faster, right?"

(Students have their own expectations and beliefs—work, study, and self-discipline will result in real reward in this class.)

"I am pleased with all of you, too. We are off to a good start! Okay, we are going to practice the basic facts of subtraction for 5 minutes, as we did in addition. You finished the addition in two days! Let's see how fast we can do the subtraction with 100 percent accuracy."

1. Practice in basic facts was done both orally and in written form (subtraction of two digit numbers). Again, it was paced orally so finger counting was not feasible.

2. Papers were corrected immediately after the test and each student corrected his or her paper. All error combinations were noted on the back of the student's paper to be studied as before.

"You will each keep a record of your scores in your folders like you did in addition. At the end of the week, you will make another *graph* of your progress in subtraction."

"I like keeping track of my own grades, Mrs. Yamate. Then, I know how I am doing and my report card grade won't be a shock. It also reminds me that if I didn't do so good one day, I know I have to pick myself up the next day," said Douglas.

(Students are showing signs of taking responsibility for their own learning.) On the third day after we started the basic facts in subtraction, I wrote on the chalkboard:

CONGRATULATIONS! You all obtained 100% on yesterday's test.

On entering the classroom, the students looked at the chalkboard and yelled,

"All right!" and slapped each other's hand with pride.

"Mrs. Yamate, you told us that we could sit with our friends when we become responsible. Don't you think that we're responsible now for our behavior? No one has received a detention for two weeks and we're doing our work," said Bryan.

"Yes, you've proven to me that you are responsible and cooperative. Okay, let's relocate ourselves! Just remember that the person you sit next to will become your partner when I say, 'Today, you will work with your partner.' You know what? Sometimes a person whom you think is a 'nerd' may be good in math, and when you get to know him or her, you may find that person to be pretty neat. End of lecture."

(Important to keep promise and reward the students for being responsible for their behavior.)

(Some students chose to sit next to a friend, some grouped in three, and one student chose to sit by himself. He said, "...so no one can bother me when I'm thinking.")

The following examples were written on the board to solve:

a) 295
    -43

b) 835
    -287

c) 7002
    -2471

d) 4000
    -1357

"Class, what is your answer for the first problem?"

"One group responded '252'; another group said, 'No, it ain't. It's 2412.'"

"Whoever obtained the answer '2412' please come to the chalkboard and show us your work."

"Three of us came up with the same answers, Mrs. Yamate. Hey, Darryl, why don't you go up and do it on the chalkboard for our team?" said Charles. Darryl said out loud: "Well, I borrowed a 10 from the 9, so I have 15. Fifteen subtract 3 is 12. Eight subtract 4 is 4, and 2 subtract nothing is 2."

295
-43

Darryl's work:

$$2815$$
$$-43$$
$$2412$$

"I borrowed like last time," said Darryl.

"Darryl, can you take away 3 from 5?" asked Selina.

"Yeah," said Darryl.

"Can you take away 40 from 90?" Selina continued. Jennifer and I got 252. Can I come up and show you?"

"Sure, if it's wrong," said Darryl.

Selina wrote on the chalkboard:

2 9 5	I had
- 4 3	I gave
2 5 2	I have left

"Try putting a cloud around the top number to remind yourself like Mrs. Yamate does," said Selina. "Okay, 4 3 *from* 2 9 5 is what? or you can think 2 9 5 minus 4 3 equals what? Then you don't have to borrow."

"Yeah, I got 2 5 2 now. That's right, right?" asked Darryl.

(The students are all eager to help each other. They are not embarrassed to come to the chalkboard and share their work. The class clapped for Darryl.)

"Yes. Thank our student teacher Selina, Darryl."

"Thanks, sister!" said Darryl. "I saw this kind of problem in the test. I did-n't know which number to put on top cause it said something like 'Subtract one hundred two from one hundred thirty.'"

"Time out! I would like to go over your use of the word 'borrowed.' You 'borrowed a '1' to make a '15' in the first column?"

"That's the way I learned it. Right, class?" said Darryl.

"Yeah. You go to your next door neighbor and borrow a '1' if you don't have enough," said Matt.

"Please listen carefully. I wrote on the chalkboard the following:

4 9 2	Expanded notation:	4 0 0 + 9 0 + 2 (minuend)
- 4 3		- 4 0 + 3 (subtrahend)
		4 0 0 + 8 0 + 1 2 (rewritten minuend)
		- 4 0 + 3 (subtrahend)

You are not borrowing a '1' from the '9.' Your previous teacher may have told you to 'borrow a '1' from the '9' and add it to the next right number. The teacher may have taught you a shortcut method by having you think, or use each digit as '1.' It is important that you understand the process of regrouping a unit of higher value into unit of lower value in order to subtract.

Then, you can use the shortcut method. Remember, our decimal system uses the ten digits, 0, 1, 2, 3, 4, 5, 6, 7, 8, and 9. A group of ten units are needed to make one unit in the next place to the left. The value of each place is a power of ten (10,100, 1000, etc.). Please study the Place Value chart in your folders, if you forgot it."

"Who would like to come to the chalkboard and do the next problem?"

"We got 4 5 8," said Fred. "I'll come up for me and my partner, Doug."

"I can't take 7 away from 5 and 8 away from 3, so I borrowed two 1's from 8:

$$
\begin{array}{r}
6 \\
7 \\
\end{array}
$$

Fred's work: 8 ^{1}3 ^{1}5   (minuend)

     - 2 8 7 (subtrahend)

        4 5 8   (remainder)

The following was written on the blackboard:

Expanded Notation

   8 3 5 = 8 hundreds 3 tens 5 ones = 7 hundreds 12 tens 15 ones

-    2 8 7 = 2 hundreds 8 tens 7 ones = 2 hundreds 8 tens   7 ones

                                                   5 hundreds 4 tens   8 ones

        8 3 5 = 8 0 0 + 3 0 + 5   Minuend
    -   2 8 7 = 2 0 0 + 8 0 + 7   Subtrahend
                                               Remainder

        8 3 5 = 7 0 0 + 1 2 0 + 1 5   Renamed Minuend
    -   2 8 7 = 2 0 0 +   8 0 + 7   Subtrahend
                500 +   4 0 + 8 = 5 4 8   Remainder

Shortcut Method:

"Let's go over Fred and Doug's work, using the shortcut way:

```
                 2  15           7  12  15
     8 3 5       8 3 5          8  3  5      (minuend)
   - 2 8 7     - 2 8 7         - 2  8  7     (subtrahend)
                               5  4  8       (remainder)
```

"Let's do this problem together. And please add these words to your vocabulary list under *Notes*: Minuend; Subtrahend; Remainder or Difference.

First, we will decompose the 30 into 2 tens and 10 ones and add the 10 ones with the 5 ones. Now, we can take away 7 ones from 15 ones. The difference is 8, so we place the 8 in the ones column," the students replied. Mrs. Yamate, let *us* finish this now with our partners. We get it," said Jason.

Jason and Neal said, "We can't take 8 tens from 2 tens, so we break up the 8 hundreds to 7 hundreds and 10 tens. Add 10 tens to 2 tens and get 12 tens. Now, we can take 8 tens away from 12 tens. We have 4 tens left."

"How many hundreds do we have left, students?"

"Seven hundreds. Now we can take away 2 hundreds from 7 hundreds. We have 5 hundreds. The answer is five hundred forty-eight. Gotcha, Mrs. Yamate. This explains everything now!" said Jason. "Boy, are we smart!"

"Kenya, did you find your error?" asked Neal.

"Yeah, I thought you had to go to the 8 and borrow for both 3 and 5," said Kenya. "How come no one taught us like this before?"

(Possibly, diagnosis of error was not done previously, or correction was not made. Often time during classroom correction, an individual student's error may not be noted, only the end result—score.)

"There's another helpful method called the 'Hutchings Low-stress method' (Hutching, B., 1975).

```
       8   3   5        Minuend
       7  12  15        Renamed minuend
   -   2   8   7        Subtrahend
       5   4   8        Remainder
```

The minuend (sum) is renamed with the help of half-space digits, and the renamed minuend is written between the given minuend and the subtrahend (known addend).

You may find that your friend is using a different method. There is no one way of arriving at a solution. I want you to find a method that is easiest for *you*."

"Everyone, please check your answers by adding the remainder with the subtrahend."

"What's a subtrahend? I think I saw that word on a problem in the test," said Jimmy. "But I never learned it, so I couldn't do it."

"You know what a *SUB*marine is? Well, then the *sub*trahend must be which number—the top or the bottom number?"

"Well, a submarine goes under the water. Underneath. Then the top number must be called the *minu*end, right?" said the class in unison.

"Wait a minute, Mrs. Yamate. I didn't get the top number when I added the subtrahend and the difference, wait for us," said Jason and Neal.

"Okay, ready to continue? Any questions? Okay, what do you have for the next problem, class?"

"5471"…"No, 4531." (Different answers from different group.)

"Latania, will you come to the blackboard and show us how you got 5471?"

"We got 5471, too," said Derrick and Jimmy.

Latania's work:
$$\begin{array}{r} 7\,0\,0\,2 \\ -\,2\,4\,7\,1 \\ \hline 5\,4\,7\,1 \end{array}$$

Latania started, "Well, 2 subtract 1 is 1, 7 subtract 0 is 7, 4 subtract 0 is 4, and 7 subtract 2 is 5. So it's 5471."

Andrew and Michael said, "We got 4531. Come on, can we go up and show you guys how to do it? We know where you made the mistake."

"Yeah," said Latania.

"First, put a cloud around the top numbers. Now, you do the problem again," said Andrew.

$$\begin{array}{r} {}^{6}\ {}^{9} \\ 7\ 0\ {}^{1}0\ 2 \\ -\ 2\ 4\ 7\ 1 \\ \hline 4\ 5\ 3\ 1 \end{array} \qquad \begin{array}{r} 6000+900+100\ +2 \\ -\ 2000+400+\ 70\ +1 \\ \hline 4000+500+\ \ 30\ +1=4531. \end{array}$$

(When students made errors in their work, they were willing to accept their classmates' help, and also give help. The feeling of 'We are *all* in this together,' which was stressed at the beginning of school year, has developed a cooperative, positive learning group. Students remind each other of their class goal: To get to the green book and red book. (The students know that I am willing to teach them out of the textbook used by the 'brainy' students—not available for use in this class.)

"Two subtract 1 is 1. Oh, I have to borrow…can't take 7 away from 0, can't borrow from the next number either because it's 0, so I have to borrow from the 7 thousand. Now I have 6 thousand left, I have 10 hundred so I can lend 1 hundred, now I have 9 hundred left. Now I have 10 tens so I can subtract 7 tens. Okay, I got 4531," said Latania.

(The class clapped. Whenever a student solved a difficult problem, the class would clap. Peer appreciation.)

"Class, what did you get for the next problem?"

"We got 2 6 4 2," said one group; "No, it's 2 6 4 3," answered another group.

"Rodney, will you come to the board and show us how you got 2642?"

"Sure, it's easy," said Rodney.

$$
\begin{array}{ll}
\text{Problem:} \quad
\begin{array}{r}
4\ 0\ 0\ 0 \\
-\ 1\ 3\ 5\ 7 \\
\hline
\end{array}
&
\text{Rodney's work:}\quad
\begin{array}{r}
3\ 9\ 9\ 9 \\
-\ 1\ 3\ 5\ 7 \\
\hline
2\ 6\ 4\ 2 \\
\end{array}
\end{array}
$$

"Hey, you can't change all the zeros to nines," said Walter.

"Yep, that's the way I learned it. My last teacher told me that when I have lots of zeros, I can change the first number to one less and knock off all the zeros and make them nines," replied Rodney.

"Can I go up and show you where you made the mistake?" asked Walter.

"Yeah, but I know I'm right because my partner got the same answer," responded Rodney.

$$
\begin{array}{r}
4\ \ 0\ \ 0\ \ 0 \\
3\ \ ^{1}0\ \ ^{1}0\ \ ^{1}0 \\
\hline
3\ \ 9\ \ 9\ \ ^{1}0 \\
\end{array}
\qquad\qquad
\begin{array}{r}
3\ 9\ 9\ ^{1}0 \\
-\ 1\ 3\ 5\ 7 \\
\hline
2\ 6\ 4\ 3 \\
\end{array}
$$

"Okay, I get it. So the last number is a 10 now. I got 2 6 4 3. Thanks, brother!" said Rodney.

"Will you all write these problems onto your *Notes*, please?

Your homework assignment will cover all these types of problems. We need more practice. Now, we must check our answers. What operation would we use to check and why?"

"Addition, because it's the opposite of subtraction!" chorused the class.

"You add the difference and the subtrahend, and if your answer is correct you will get the same minuend number," said Joanne.

"Wow, that's great! You're learning the terms used in subtraction like a mathematician!"

$$
\begin{array}{rl}
130807 & \text{(minuend)} \\
-\ \ \ 95876 & \text{(subtrahend)} \\
\hline
& \text{(remainder or difference)}
\end{array}
$$

"Let's do this problem together. Did you all get 34931 as the remainder?"

$$
\begin{array}{rl}
130807 & \text{(minuend)} \\
-\ \ \ 95876 & \text{(subtrahend)} \\
\hline
34931 & \text{(difference)}
\end{array}
\qquad
\begin{array}{rl}
95876 & \text{(addend)} \\
+\ \ 34931 & \text{(addend)} \\
\hline
130807 & \text{(sum)}
\end{array}
$$

"Why did we get the same minuend number when we checked our solution?"

"Because subtraction is the *inverse* or the opposite operation of addition. That's why we call the difference a sum and the subtrahend and minuend as addends when we go in the opposite direction. Now we get it," the class chorused.

"Yeah, now I see why my teacher kept calling it the sum," said Michael.

Skill Practice (noting specific problems).

Examples in which the number in the minuend is less than the number in the subtrahend.

8 7 2 - 4 9	Regrouping in ones' and tens' places.
7 3 4 - 3 7 6	Regrouping in ones' and tens' places.
5 3 6 - 9 7	Blank place in subtrahend; regrouping in hundreds', tens', and ones' places.
8 0 0 - 4 8 9	Double zeros in minuend requiring regrouping in hundreds', tens' and ones' places.
3 0 8 - 2 9	Zero in tens' place; regrouping in hundreds', tens', and ones' places.
9 0 0 - 3 0 8	Double zeros in minuend requiring regrouping in all places.

The following day

"Today, we will take a test on addition and subtraction that we've been studying. After you finish your test, I would like you to do the following:

Problems 1, 3, 5, and 8–12 on page 235 in your text."

Mike finished his test first. He went by Andrew's desk and asked, "What did she tell us to do after the test?"

(Michael always jumps into an activity immediately and fails to listen to the rest of the directions. He becomes anxious as soon as an   assignment is verbal. He has a difficult time sitting or concentrating for 10-15 minutes. I have given him permission to walk about in the classroom if he does not disturb other students.)

"Look on the chalkboard, Mike. You know Mrs. Yamate always writes what she tells us to do," said Andrew.

The assignment on the chalkboard: Today we will study text p. 235.

Please do problems 1, 3, 5, 11-15.

(Michael went up to the chalkboard and stood sideways, writing something on the chalkboard. The students started smiling and looked towards me. He finished and went back to his seat. He had changed the assignment to read:

Today we will study <u>sex</u> 235.

The students all turned their heads toward me. I laughed and the whole class laughed!)

I wrote on the chalkboard:

Number word	Latin Root	Numerical Equivalent of Root
Billion	Bi-	2
Trillion	Tri-	3
Quadrillion	Quater	4
Quintillion	Quintus	5
*Sex*tillion	*Sex*	6

"Class, please memorize the above to 'Sextillion' now! "Your Health Education teacher will help you with the rest."

(The class is easing up and doing humorous antics like any teen-agers! I took the opportunity to expand their numeration.)

Another day

(Jessie walked in front of John's desk after sharpening his pencil on the front left corner wall. As he was returning to his desk, he wrote across John's desk with the pencil.)

"Brother," said John out loud.

"Sister," retorted Jessie.

"Mother," said John.

(Jessie, who had gone back to his seat by this time, got up and went back towards John's desk.)

"Students, please put your pencils down. I need to have you tell me what this brother, sister, mother, is all about. I have a brother, a sister, and a mother. Telling me that wouldn't get me so mad as Jessie and John are now."

(I had never heard of such talk between students before. Now, the *teacher* is the *learner*.)

Charles whispered, "Well, Mrs. Yamate, if someone says, 'Your ma ma' like this (intonation), boy, there is really going to be a fight!"

"Why is it so bad to say 'Your ma ma'?"

"It's just the way it is. I don't know why," said Nika.

"Well, no one's going to talk about my mother!" said Jessie with a glare.

"It's like calling your mother a prostitute," said Charles softly.

"Class, let's watch what we say to each other. There's enough garbage out there in the outside world without our throwing it about in our class. What happened to your promise? What happened to your class goal? You all set it at the beginning of school. What happened to the respect and care for each other? We are classmates and friends in this class. Apologies are in order, John and Jessie. I am really disappointed in your behavior!"

"Sorry," the boys said to each other. "We don't want to disappoint you, Mrs. Yamate. It won't happen again, okay."

The following day

(I noticed that Ned had his head down on his desk.)

"Ned, are you sleepy? I noticed that your head has been down on your desk all morning. Are you all right?"

"I've had this headache all week-end, Mrs. Yamate. No headache pills have stopped the pain," said Ned.

"The school nurse only comes on Wednesdays, so you'd better tell your mother about your continuing headache. You need to be checked by a doctor."

(Ned has been out of school for a whole week now. One of the students in the class announced that Ned had brain surgery. Upon checking with the school counselor and the hospital, we were advised that we could go and visit him. The principal would not give me permission to take four students to the hospital after school. They would have to go with their parent or guardian. I had each student tape a message for Ned and took it to the hospital. I also brought Ned our new math textbook to keep him in contact with his class. [The four students went by bus to the hospital by themselves.])

(When I entered Ned's room, his mother was sitting by the window and greeted me aloofly. Ned was no longer the healthy looking student I had.

He had lost his hair due to chemotherapy. He had lost a lot of weight. He looked like a malnourished child. He was now blind due to malignant brain tumor.)

"Hello, Ned."

He immediately recognized my voice and said, "Mrs. Yamate, you came at last!" I hugged him and he hugged me back with tears in his eyes.

I brought you a tape recorder and messages from your classmates, Ned.

"Oh, that's Naomah; Oh, that's John; Oh, that's Walter."

(He recognized everyone's voice.)

"I brought you a new textbook that we are using now. I forgot it in the car, so I will go and get it for you."

(When I returned, Ned was sitting up in his bed.)

His mother said, "This is the first time I've seen Ned sitting up. Since you left the room, he has been sitting up listening for your footsteps. He must be really happy to see you."

(I gave Ned the new math book and read over the topics to him. He was good in math, so I knew he would be happy to receive this book.)

"I'll study so I won't get behind the class," said Ned.

"I'll come and see you again, Ned. Next time, I'll go over our class work with you."

(I want Ned to feel connected to his classmates and give him hope to get well, so he could return to school.)

Two months later I found a message in my school mailbox. I had transferred to another middle school. "Please call Ned at 213-4856."

I called and found Ned at home. I told him that I would come and see him as soon as I get caught up in my work.

Another message followed a week later: "Please contact Ned at once.

It is urgent."

I went to see Ned at his home. His cousin was taking care of him while his mother went to work.)

"Mrs. Yamate, you finally came to see me! I don't have too much time left. I wanted to see you so badly. I couldn't find you because you transferred to another school, and the office wouldn't give me your new school number or which school you transferred to. Finally, Mrs. W. who was my counselor told me she would give you my message," said Ned. (We hugged each other and cried.)

Ned is now blind. He said, "See those pictures on the mantle? They are pictures of my family and me. My father and mother are divorced, and my father has two children by his second wife. I am my mother's only child. I live here with my mother. I am worried that if I die, my mother is going to be all alone. (A dying 13 years old worried about his mother whom he will have to leave behind.) I am so happy that you came to see me, Mrs. Yamate."

"Ned, I am sure that your aunt and uncle will care for your mother. Just take care of yourself at this time."

"Thanks for the tape recorder and the tapes. I can listen but I can't see any more. Do you know what a last wish is?" asked Ned.

"Yes."

Ned said, "I'm going to Disney World Saturday, so I wanted you to come before I leave just in case I don't make it."

"I'm glad I was able to come today, then, Ned. You'll have a great time at Disney World. I am sorry that I did not come to see you sooner."

(I hugged him good bye and tears rolled down our faces.)

"Thank you, Mrs. Yamate, you are my favorite teacher. Thanks for coming to see me," said Ned.

(Ned died a three weeks later.)

LIST OF COMMON ERRORS:

1. Computation error.

2. Regrouping error.

3. Subtracted minuend from subtrahend when the minuend was a smaller

number: 6 4 3
              - 2 8 9
                 4 4 6

4. Errors due to zeros in minuend: 4 0 5 0   5 0 0 0
                                         - 2 7 5 9   - 2 8 9 7
                                            2 3 0 1   3 8 9 7

5 . Regrouped the minuend unnecessarily: 5 7 9 ^{1}8
                                             - 2 3 3 6
                                                3 4 5 12

6. Increased minuend number after decomposing: 5 4 2 0
                                                 - 3 2 1 7
                                                    2 2 2 3

7. Decreased hundreds twice - problem with zeros: 7
                                                       8
                                                    9 0 0
                                                 - 2 8 8
                                                    5 1 2

CHAPTER 3

OBJECTIVE 3: MULTIPLICATION OF WHOLE NUMBERS.

This objective requires students to solve complex multiplication—processes of multiplying by two or more numbers, including ending in zeros or including zeros.

TEST ITEMS:

a) $\begin{array}{r} 829 \\ \times\ \ \ 17 \\ \hline \end{array}$

b) $\begin{array}{r} 9671 \\ \times\ \ \ \ 27 \\ \hline \end{array}$

c) $\begin{array}{r} 4000 \\ \times\ \ \ \ 30 \\ \hline \end{array}$

d) $\begin{array}{r} 9080 \\ \times\ \ \ \ 89 \\ \hline \end{array}$

e) 396 x 42

BASIC KNOWLEDGE NEEDED

1. Know multiplication tables through 9 x 9.

2. The meaning of the terms: multiplication, factor, product, multiplicand, multiplier, carrying, and sum.

```
   2 4 6   Multiplicand          2 4 6  x   6  =  1 4 7 6
 x     1 6   Multiplier          (factor)    factor)  (product)
   1 4 7 6
 + 2 4 6     Partial Product
   3 9 3 6   Product
```

3. Multiplication is the process of repeatedly adding a quantity to itself a certain number of times. The *multiplicand* is the number to be added to itself. The *multiplier* is the number that tells how many times this addition is to be done.

4. The ones must be placed under the ones, tens under the tens, hundreds under the hundreds, etc.

5. Zero "0" is a place holder.

6. Ability to multiply and then add the number carried.

7. Ability to handle a "zero" or a succession of "zeros" in the multiplier.

8. Ability to use the multiplier with the tens or hundreds number in the multiplicand and where to place the partial products.

9. Ability to add these partial products.

10. Know how to check for correct answers.

EFFECTIVE STRATEGY—Mastery of The Basic Facts in Multiplication (while carrying on dialogue with students).

"Students, we are going to check our basic facts in multiplication today.

I will time you and when I say stop, you will mark a line across your paper. Then, you will finish the rest of your paper as you did before in addition and subtraction. We are going to do this every day for 5 minutes until we get 100 percent correct. You will keep a record of your scores in your folders. At the end of the week, we will make another *graph* of your progress."

(Clarity of what is to be learned. Pacing to meet all students' time frame.)

"Quick, quick, someone, tell me what's 8 x 7. I always forget," said Antarra.

"Whenever you see a 7 and an 8, 5 and 6 comes along, so it is 56."

"Gee, that's a neat way of remembering," said Antarra. "I won't forget now!"

1. Practice in basic facts was done both orally and in written form. It was orally paced so that the students did not have the time to add multiples of one number to obtain the product.

2. Papers were corrected immediately after the test and each student corrected his or her paper. All error combinations were noted on the back of the student's paper to be studied as before.

"You will each keep a record of your scores in your folders like you did in addition and subtraction. At the end of the week, you will make another graph of your progress."

"Darwin, why are you talking?"

"How can you tell, Mrs. Yamate. My back is towards you," said Darwin.

"When you talk too much your mouth starts to protrude from the other side of your head so that the teacher can see."

"Oh please, don't give me a detention (He pleaded kneeling to the floor in prayer). Next time you catch me talking, you can throw me out the window," said Darwin.

"No thanks, Darwin. I don't want to carry you out the window...besides, you are bigger than I am!"

(The class laughed out and said, "Yeah, three times bigger!")

(I laughed and shook my head. How can I give him a detention with such a plea?) "So, it's a warning!"

(Could these students have been such behavior problems that the teacher could not teach them before?)

"Patrick, why are you tilting your desk?"

"I can think better," said Patrick.

"I don't want you to fall backwards in my class. Besides, I don't want to pick up the pieces of your body." (Patrick smiled.)

(The seat is attached to the desk, so most of the students find it uncomfortable for their large bodies. They can lift up the desk with their thighs, as is the case with Patrick.)

Skill Practice (noting specific problems)

a)
$$\begin{array}{r} 3010 \\ \times5 \\ \hline \end{array}$$

d)
$$\begin{array}{r} 312 \\ \times400 \\ \hline \end{array}$$

b)
$$\begin{array}{r} 804 \\ \times7 \\ \hline \end{array}$$

e)
$$\begin{array}{r} 369 \\ \times78 \\ \hline \end{array}$$

c)
$$\begin{array}{r} 612 \\ \times30 \\ \hline \end{array}$$

f)
$$\begin{array}{r} 596 \\ \times143 \\ \hline \end{array}$$

"Let's do these problems together, class. Who would like to come to the chalkboard and do the first problem?"

(All students are eager to come to the chalkboard. They know that it is not a punishment or to show off. The students are sharing and learning from each other and developing a friendship and respect for each other's skill. Mark was first to the chalkboard and solved the problem.)

"Very good, Mark. Did you all get 15,050?"

"Yes," responded the whole class."

"Great! Who would like to do the next problem?"

(Many hands go up. Gregg comes up to the chalkboard.)

$$
\begin{array}{r}
\overset{2}{8\ 0\ 4} \\
\times\ \ \ \ 7 \\
\hline
5\ 6\ 2\ 8
\end{array}
\qquad
\begin{array}{r}
8\ 0\ 4 \\
\times\ \ \ \ 7 \\
\hline
5\ 6\ 2\ 8
\end{array}
$$

"I like the way you circled the number you carried to remind yourself. Since you know that 7 x 4 is 28 and 7 x 0 is 0, you can put the 28 down in the answer without showing that you carried the 2 in the tens column. Hold the 20 in mind. This will save you time on a test."

The following examples were written on the chalkboard to solve:

$$
\begin{array}{r}
6\ 1\ 2 \\
\times\ \ \ 3\ 0 \\
\hline
\end{array}
\qquad
\begin{array}{r}
3\ 1\ 2 \\
\times\ 4\ 0\ 0 \\
\hline
\end{array}
$$

"The first answer is 1836!" said Tyshenna.

(Tyshenna came up to the board to do the problem.)

$$
\begin{array}{r}
6\ 1\ 2 \\
\times\ \ \ 3\ 0 \\
\hline
1\ 8\ 3\ 6
\end{array}
$$

"Wait a minute, Tyshenna. What happened to the zero in the multiplier?"

"Well, it's zero so I don't need to multiply it. Anything multiplied by zero is zero," replied Tyshenna confidently. "Yeah, 0 x 2 is 0; 3 x 1 is 3 so I put the 3 down. 3 x 6 is 18, so I put the 18 down."

"Do I have to remind you of our place value system again? The value of 0 in 30 is 0 ones, but the value of 3 in 30 is 3 times 10 or 30 ones, or 3 tens."

"That's why you need to put the zero down first. The multiplier is 30 and not 3," said Samuel. "How many times does Mrs. Yamate have to tell you?"

"How many of you put down three zeros in the first row?"

(Everyone raised their hands.)

```
    6 1 2              6 1 2
  x   3 0            x   3 0
    0 0 0              6 0   (3 0 x 2)
+1 8 3 6              3 0 0   (3 0 x 1 0)
  1 8 3 6 0        +1 8 0 0 0   (3 0 x 6 0 0)
                     1 8 3 6 0
```

"You can save yourself time in a test if you would write a zero under the ones digit in the partial product (put an arrow under it) and eliminate rest of the zeros in the first row of partial product and continue to multiply the rest of the digits by 3 tens."

```
    6 1 2   (multiplicand)
  x   3 0   (multiplier)
  1 8 3 6 0   (product)
```

"Eric, would you like to do the next problem?"

"Okay," said Eric and proceeded:
```
      3 1 2
    x 4 0 0
      0 0 0
    0 0 0 0
  1 2 4 8 0 0
  1 2 4 8 0 0
```

"Wow, that's a lot of zeros in your problem, Eric. That's going to take up much of your time when you take a timed test—like the achievement test. Let's see how we can shorten it.

```
    3 1 2   (multiplicand)
  x 4 0 0   (multiplier)
  1 2 4, 8 0 0   (product)
```

Put an arrow under the ones, and arrow under the tens, and an arrow under the hundreds, then you won't lose your place. Also, when you have zeros in the multiplier, you can multiply the multiplicand by the numeral in the multiplier (4 hundred) and add the number of zeros you have at the end as a short cut."

```
        3 1 2  (multiplicand)
     x  4 0 0  (multiplier)
        0 0 0
      0 0 0 x  (partial product)
    1 2 4 8 x x
    1 2 4 8 0 0  (product)
```

"I see that some of you put an 'x' as a place holder. A 'x' is meaningless here: 3 1 2 x 4 0 0 is not 1 2 4 8 x x. *Zero* is a number and is essential to the value of each place in our decimal number system. For example, the difference between '1' and '10' is the addition of '0'—changes the value."

"But, that's the way our last teacher taught us. Put 'x' in all the empty spaces," said Imoni.

"Let's work the next problem together; add the mathematical terms to your vocabulary list, please."

```
        4  6
        5  7

        3  6  9     (Multiplicand)
     x        7  8  (Multiplier)
        2  9  5  2  (Partial Product)
   +  2  5  8  3
     2  8  7  8  2  (Product)
```

"Some of you may find it easier to multiply this way:

```
     3 6 9              3 6 9
   x     8            x     7 0
   2 9 5 2     +      2 5 8 3 0    =    2 8 7 8 2
```

Now that you all know your basic facts, it won't be difficult to do multiplication problems, right?"

"Mrs. Yamate, I didn't get 2952 in the first row. I got 10442," said Sasha. "I did 9 x 8 = 72, carry the 7 and put down the 2; add 7 + 6 = 13, 13 x 8 is 104; put down the 4 and carry the 10; 10 + 3 =13; now, 13 x 8 is 104."

$$
\begin{array}{r}
{}^{10}\ \ {}^{7} \\
3\ 6\ 9 \\
\times\ \ \ \ \ 8 \\
\hline
1\ 0\ 4\ 4\ 2
\end{array}
$$

Sasha's work:

"Let's say each step out loud:

$$
\begin{array}{r}
{}^{4}\ \ {}^{6} \\
{}^{5}\ \ {}^{7} \\
3\ 6\ 9 \\
\times\ \ 7\ 8 \\
\hline
2\ 9\ 5\ 2 \\
+\ \ \ 5\ 8\ 3 \\
\hline
2\ 8,7\ 8\ 2
\end{array}
$$

(8 x 3 6 9)
(7 0 x 3 6 9)

"You multiply 9 x 8 = 72; write down the 2 in the ones column and carry the 7 to the tens column. Next you multiply 8 x 6 = 48, now you add the 7 you carried—48 + 7 = 55. Write down the 5 in the tens column and carry the other 5 to the hundreds column. Now, 8 x 3= 24 + the 5 you carried = 29. Write down the 29. Do you understand now how we got 2952? This is a short cut method so each digit is used as 'ones.' So long as you understand the place value of each digit, you may use this method for expediency."

"That's why I've been making all those mistakes! I've been adding the carried number first, then multiplying. Okay, I get it now. I can see where I have been making the mistake *all this time*! It's a good thing you're teaching us and explaining each step for us because some of us have been doing it wrong all this time! How come other teachers didn't correct me?" said Sasha.

"Yeah, we are finally learning and understanding all this stuff," said Antone.

"I told you when you entered my class that you can ask me any questions. This is why I asked all of you to work out all the problems in the achievement test. I can now correct any misconceptions or mistakes that are fixed in your 'memory bank.' This review work, corrective work, has been really a good learning experience for *me*, too. I have been learning a lot from all of you!

I had to learn to really slow down in explaining some concept or procedure to you so you can follow me; wait for you to sort things out in your mind; find different ways to teach you depending on your different style of learning, input of information, and most importantly, making math meaningful and understandable so you can appreciate and enjoy it. It does make sense now, right? What a challenge I have because we aren't through yet! Thank you, class, for teaching me so much!" (Teacher as a learner...psychology of learning and teaching mathematics is really important!)

"Oh, it's okay. We like teaching you! Right, class? said Gentry.

"Now, these steps may help you to understand the process of multiplying 369 by 78."

Multiplying from right to left:

```
      3  6  9                      3  6  9
  x         8                  x         7  0
         7  2   (8 x 9)               6  3  0   (7 0 x 9)
      4  8  0   (8 x 6 0)          4  2  0  0   (7 0 x 6 0)
  + 2  4  0  0   (8 x 3 0 0)   + 2  1  0  0  0   (7 0 x 3 0 0)
    2  9  5  2                   2  5  8  3  0
```

```
      2  9  5  2   (8 x 3 6 9)
  + 2  5  8  3  0   (7 0 x 3 6 9)
    2  8, 7  8  2
```

"Class, what did you all get for the last problem?"

"I got 4768," said Tamesha.

"No, We got 85228," said Walter.

"We seem to have different answers for the last problem."

"I got 4768, too. Tamesha and I worked it together," said Jennifer. This is how we did it:

```
      5  9  6
  x   1  4  3
   1  7  8  8
   2  3  8  4
      5  9  6
   4  7  6  8
```

"You forgot that you were multiplying by 40 and not 4; multiplying by 100, not 1," said Samuel.

"Remember, our decimal system of representing numbers is dependent on the digits, the placement of these digits, and the value of each place. The value of 3 in 143 is 3 units, the value of 4 is 4 times 10, and the value of 1 is 1 times

100. Let's use a *graph* paper again so we can line up our numbers in the right column. Here are two methods you can use to solve this problem. See which is easier for *you* to understand:

```
        3   2
        2   1
        5   9   6
    x   1   4   3
    1   7   8   8   (3 x 5 9 6)
2   3   8   4       (4 0 x 5 9 6)
5   9   6           (1 0 0 x 5 9 6)
8   5   2   2   8   (product)
```

```
OR              5   9   6
            x   1   4   3
        5   9   6   0   0   (1 0 0 x 5 9 6)
        2   3   8   4   0   (4 0 x 5 9 6)
    +       1   7   8   8   (3 x 5 9 6)
        8   5   2   2   8
```

Skill practice (noting specific problems).

Examples in multiplication involving carrying:

```
    1 7
x     6
```
Carrying to tens' place; two-place product.

```
    7 4
x     6
```
Carrying to tens' place; three-place product.

```
    3 1 7
x       7
```
Carrying to tens' and to hundreds' place.

```
    4 3 1
x       8
```
Carrying to hundreds' place.

```
    5 1 7 2
x         4
```
Carrying to alternate places.

$$\begin{array}{r} 5\ 0\ 0 \\ \text{x} \qquad 3 \\ \hline \end{array}$$ Zero at end.

$$\begin{array}{r} 4\ 0\ 5 \\ \text{x} \qquad 1 \\ \hline \end{array}$$ Zero in middle; no carrying.

$$\begin{array}{r} 3\ 0\ 7 \\ \text{x} \qquad 8 \\ \hline \end{array}$$ Middle zero; carrying to zero tens.

$$\begin{array}{r} 4\ 0\ 0\ 8 \\ \text{x} \qquad 6 \\ \hline \end{array}$$ Double zeros; carrying to tens' place.

"What operation would we use to check our work and why?"

"We never learned that," said Walter.

"What operation is the opposite of multiplication?"

"*Division* is the *opposite* of multiplication, right?" said Fred.

"Right. We will divide the product by the multiplier. Please *check* all of the skill practice problems that we just did. Although Friday is our conferencing day, I would like to move it up to Thursday this week, so anyone having questions or challenged in multiplication of multi-digit numbers can go over it with me. We will then have our test on Friday."

(Conferencing helps me to question and discuss with a student on one-to-one basis before a test. Each student is given 5 minutes and signs up on the front chalkboard. Also, I do not test students until they have a good understanding and knowledge of the week's instruction. It is important that these students gain confidence in their abilities each step of the way. Also, I am keeping in mind my goal for the class—to *permanently* elevate them from being trapped in the low track class.)

SUMMARY

LIST OF COMMON ERRORS:

1. Weakness in combinations.

2. Carrying errors (carried wrong digit).

$$\begin{array}{r} 3\ 6 \\ \times\quad 9 \\ \hline 3\ 1\ 5 \end{array}$$

3. Errors due to zero as multiplier.

$$\begin{array}{r} 2\ 8 \\ \times\ 4\ 0 \\ \hline 1\ 1\ 2 \end{array}$$

4. Error in carrying into zero.

$$\begin{array}{r} 5\ 0\ 8 \\ \times\qquad 4 \\ \hline 2\ 3\ 0\ 2 \end{array}$$

5. Forgets to carry.

$$\begin{array}{r} 4\ 6\ 2 \\ \times\qquad 7 \\ \hline 2\ 8\ 2\ 4 \end{array}$$

6. Added carried number with multiplicand and multiplied the sum.

$$\begin{array}{r} 5\ 4\ 6 \\ \times\qquad 2 \\ \hline 1\ 2\ 0\ 2 \end{array}$$

7. Error in position of partial products.

$$\begin{array}{r} 4\ 5\ 2 \\ \times\quad 3\ 2 \\ \hline 8\ 0\ 4 \\ 1\ 3\ 5\ 6 \\ \hline 2\ 1\ 6\ 0 \end{array}$$

CHAPTER 4

OBJECTIVE 4: LONG DIVISION—TWO DIGIT DIVISORS

This objective requires students to solve long division-two digit divisors, be able to name the quotients, do multiplication of the quotient involving carrying, and do the necessary subtraction.

TEST ITEMS:

a) 8 $\overline{\smash{\big)}\ 8\ 4\ 0}$

b) 5 7 $\overline{\smash{\big)}\ 2\ 4\ 5\ 1}$

c) 6 9 ÷ 3 2

d) 4 0 8 ÷ 6

e) 6 4 6 4 ÷ 8 8

LIST OF COMMON ERRORS:

1. Combination error.

2. Carrying error.

3. Error in multiplication.

4. Error in subtraction.

5. Used remainder larger than divisor.

6. Omitted zero in dividend.

7. Used long division form for short division.

8. Used dividend as divisor in horizontal division.

EFFECTIVE STRATEGY—Mastering The Basic Facts in Division (while carrying on dialogue with students).

"Students, since we mastered our basic facts in multiplication, I know that we can master our division facts. We will use the same procedure as we did in other operations. At the end of the week, you will make another <u>graph</u> of your progress. By the way, I checked the *graphs* and your *notes* in the folders. I see great progress has been made by all of you. I was also happy to read the many comments made by your mother/parent/ guardian, when they signed your folders. I have placed an extra star inside each folder."

"Remember you promised to teach us out them green and red books, Mrs. Yamate. We've *really* been studying hard and being quiet so you can teach us," said Imoni.

"And we've been helping each other too," said Charles.

"I did not forget, students. I am very proud of this class!"

1. Drill in basic facts was done both orally and in written form. It was paced so the students could not count to get quotients.
2. Papers were corrected immediately after the test and each student corrected his/her errors. All error combinations were noted on the back of his/her paper to be studied as before.

"May we sit in a circle today?" said Elton upon entering. "We're going to do this short division paper?"

(Each student picked up a worksheet and his/her folder.)

"Yes, you may."

(Reward for doing their work independently as well as in groups as requested. Now, they are expressing their desire as to how they would like to work. They feel that this is their class, their classroom now.)

"Mrs. Yamate, are there remainders?" asked John. "I studied last night, so I know I can do these."

"These are easy. Do you want to time us?" said Elton.

"All right, I'll time you. Yes, there are remainders, John."

Examples of thirty problem: $5\overline{)3\ 9}$ $8\overline{)6\ 2}$

"We're ready to move up to the next step. Ready for the challenge? I won't time you on these problems; please do them carefully."

2 | 2 6 8   Even division; three place quotient.

3 | 6 0 9   Even division; zero in quotient in middle.

4 | 8 4 0   Even division; zero in quotient at end.

3 | 6 0 0   Even division; double zero in quotient.

4 | 2 8 4   Initial trial dividend - two figures; no
                       remainders.

3 | 6 1 5   Zero in quotient - in middle; carrying.

3 | 3 9 5   Remainder.

2 | 8 0 5   Remainder; zero in middle.

4 | 2 8 6   Initial trial dividend - two
                       figures; remainder.

5 | 3 0 4   Initial trial dividend - two figures, zero in two
                       figures, zero in quotient; remainder.

3 | 9 1 9   Zero, carrying, and remainder.

6 | 3 9 5   Carrying and remainder.

"I see that you are checking the answers with each other. Good ! Would you like to read off the answers and I'll check them against mine?"

(Students are working in a circle in groups, checking and helping each other without being told. The students are the teachers now.)

"Okay. '29' for the second problem," said Darryl.

"What?" chorused the class.

Walter quickly looked at Darryl's paper and said, "Hey, you forgot to put a zero in the middle. Zero divided by 3 is zero, and you put the 0 in the quotient."

"Oh, thanks brother! said Darryl.

(On the problem 286 divided by 4, Gentry noticed that Antarra put the first quotient 7 over the 2 and 1 over the 8.)

Gentry said, "You've got the 7 in the wrong place, man. You're suppose to put the 7 over the last number you used, the 8, and you put the 1 over the 6. The remainder goes outside of the box."

"Okay, I get it now. I've been writing like this all the time. Man, that's why I've been hating divisions. I always got them wrong. Thanks, brother!" said Antarra.

"I am happy to see all of you helping each other. You know, in the business world 'cooperation' is the key word. You share knowledge with your co-workers, work together to get things right and produce, just as you all are doing now."

"Will you please check your answers by multiplying the quotient by the divisor? Please add these words to your vocabulary list:

$$\begin{array}{r} 2\ \ 0\ \ 2\,\tfrac{1}{4} \quad \text{quotient} \\ \text{divisor}\ \ 4\ |\ \overline{8\ \ 0\ \ 9} \quad \text{dividend} \end{array}$$

You will do page 290 for homework. Let's go over some of the kind of problems you will have to do."

$$40\ |\ \overline{8\ 0} \qquad 32\ |\ \overline{6\ 9} \qquad 80\ |\ \overline{5\ 6\ 0}$$

$$43\ |\ \overline{2\ 2\ 7} \qquad 63\ |\ \overline{4\ 2\ 9} \qquad 55\ |\ \overline{4\ 2\ 2}$$

"I see that some of you are stuck on 227 divided by 43. You have to estimate how many sets of 40 are there in 227? Look at the 4 in the divisor and underline it. Look at the 22 in the dividend and underline it. Now, how many sets of 4 are in 22?"

"5," chorused the class.

"Now try multiplying 43 x 5. If the remainder is not larger than the divisor, you are correct."

$$
\begin{array}{r}
5 \ ^{12}\!/^{43} \\
\hline
4\,3 \ \ | \ 2 \ \ 2 \ \ 7 \\
2 \ \ 1 \ \ 5 \\
\hline
1 \ \ 2
\end{array}
\qquad
\text{Check:}
\qquad
\begin{array}{r}
4 \ \ 3 \quad \text{(factor)} \\
\text{x} \quad 5 \\
\hline
2 \ \ 1 \ \ 5 \\
+ \ \ 1 \ \ 2 \quad \text{(remainder)} \\
\hline
2 \ \ 2 \ \ 7
\end{array}
$$

"Please copy this problem down onto your notes."

"Let's use the graph paper again to do these division problems, so we can keep track of our place.

$$
\underline{5} \ \ 7 \ \ | \ \ \underline{2} \ \ \underline{4} \ \ 5 \ \ 1
$$

"Underline the 5 in the divisor and the 24 in the dividend. You can see that 57 in the divisor is close to 60 than 50, so round up the divisor to 60 and estimate. Whenever the one's digit in the divisor is 6, 7, 8, or 9, round up—add one more to the tens digit."

(Rounding up from 6 instead of 5 gives closer estimate, although we have taught round up if 5 and above.)

$$
\begin{array}{r}
4 \ \ 3 \\
\hline
\underline{5} \ \ 7 \ | \ \ 2 \ \ 4 \ \ 5 \ \ 1 \\
- \ 2 \ \ 2 \ \ 8 \\
\hline
1 \ \ 7 \ \ 1 \\
- \ 1 \ \ 7 \ \ 1
\end{array}
$$

(*Think* 60 - for 57)   (5 7 x 4)

"60 x ? = 2 4 0"   (5 7 x 3)

"Be sure and use an arrow to bring down the next number in the dividend, then you won't get lost."

Skill Practice (noting specific problems).

Examples in long division.

$$34 \mid 108 \qquad 52 \mid 161 \qquad 35 \mid 107 \qquad 32 \mid 162$$

$$22 \mid 9350 \qquad 11 \mid 2838 \qquad 77 \mid 7854 \qquad 32 \mid 6560$$

"Mrs. Yamate, I always get mixed up when the division is written like: $408 \div 204$. I don't know which number goes inside the box," said Kenya. "Okay, let's write it on the board:

$$204 \mid 408 \qquad 408 \text{ divided by } 204$$

"Four hundred and eight (write 4 0 8) divided by (draw |) two hundred and four (write 2 0 4 outside of the division sign). Everyone, write it and say it three times, four times…as many times as you need to say it and do it in order to get it into your head, your brain… and use all your senses. .

"Yeah, because Mrs. Yamate tells us to take notes, I find that I am learning more. I am remembering the stuff she writes on the board. That's why I'm doing better on my tests," said Antarra

"Class, I find that conferencing on Thursday has become helpful for those of you who need extra help. May we change the conference time to Thursday?"

"Hey, class, don't you think it's better to change to Thursday?" said Gentry. "Let's take a vote."

(Necessary to be flexible on schedule based on students' needs.)

"All right, Mrs. Yamate. We can change to Thursday," said Gentry.

"Socializing time! Look at the clock. Everyone hurry up and pass the folders to the front so Latanesha could collect them. Michael, hurry up with the waste basket. Hey, don't throw the paper at Michael. We're going to lose our privilege," said Charles.

Gentry, Charles, Elton, and Lhyn had a new rap (communication of words in rhythm) that they wanted to do in front of the class, so Charles was anxious for the last five minutes, which is reserved for socializing. Walter and Naomah started rhythming. These students signed up for the upcoming School Talent Show.)

(Accountability for real reward based on student interests.)

SUMMARY

LIST OF COMMON ERRORS

$$3\ 1\ \overline{|\ 6\ 8\ 2}$$

No carrying in multiplication;
no borrowing in subtraction.

$$3\ 5\ \overline{|\ 7\ 7\ 0}$$

Carrying in multiplication;
 no borrowing in subtraction.

$$3\ 3\ \overline{|\ 7\ 0\ 6}$$

Borrowing in subtraction;
no carrying in multiplication;
remainder.

$$3\ 6\ \overline{|\ 1\ 1\ 1\ 6}$$

Carrying in multiplication;
borrowing in subtraction

CHAPTER 5

OBJECTIVE 5: OMPLEX ADDITION WITH DECIMALS

This objective requires students to solve complex column addition—bridging from one decade to the next.

TEST ITEMS

a)
```
        0. 6  2  5
     3  1
+       4. 5  8
```

b) 1 9 2. 3 6 + 7 3. 9 0

c) 0. 5 1 4 + 3. 4 7 8

d) 8. 2 5 + 3 9 6

e) 0. 0 5 + 0. 4 6 0 5 + 0. 4 3 6

LIST OF COMMON ERRORS:

1. Computation error.
2. Carrying errors (carried wrong numbers).
3. Incorrect placement of the decimal point in the sum.
4. Omission of decimal point.
5. Numbers were not placed correctly.
6. Whole number as decimals.

EFFECTIVE STRATEGY—Mastering The Addition of Decimals (while
carrying on dialogue with students.)

Place Value and Renaming (Carrying) with Decimals

"Please take out the place value chart that you have in your folders.
Remember, the ones, tens, hundreds, etc. We are going to add onto it going in
the opposite direction."

"A decimal point is used to separate the whole number part from the fractional part. Please note the endings that I underlined. Six tenths (.6) or 0.6 and
6/10 are numerals which name the same number."

"Why did you add a zero in front of the decimal point?" Is .6 or 0.6 right?"
asked Eric.

"We add a zero in front of the decimal point for clarification. Both are correct. Scientists and engineers always use a zero in front of a decimal point for
accuracy. It's a good habit to get into."

"I can't copy that stuff off the chalkboard. Samuel's chart is different from
mine," said Marvin.

(Marvin has difficulty copying anything from the chalkboard although he
sits in the first seat in front of the board. He appears to have spatial problem —
difficulty in remembering what was seen, copying work from the chalkboard
or textbook.)

"Please write a decimal point, Marvin, and write 'and' on top of it; write '9'
next to it and write tenths on top of it...(I continued the chart orally with him)."

"Using your chart, please write each of the following numerals as a decimal numeral:

1. Four ten*ths*

2. Five and six ten*ths*

3. Eight hundred*ths*

4. Twenty-five hundred*ths*

5. Six thousand*ths*

6. Eighty-four thousand*ths*

7. Four hundred twenty-six thousand*ths*

8. Three million*ths*

"Please work with your partners; say it to each other and write down the
numerals, and check your spelling. I will write the answers on the chalkboard
when you finish."

The following problems were written on the board to be solved:

a)   6. 4
    8. 3
    3. 7
+ 8. 1

b)   2. 8 6
   . 7
+ . 2 4

c)   8
 + . 0 5

d)   0. 8
   0. 6
   0. 3
+ 0. 4

e) 0. 5 2 3 + 3. 1 8 + 1 2 + 2. 2

"Let's check our work. Good, I see that you all got problems 'a' and 'b' right. Some of you put a '0' in the vacant space next to .7 in problem 'b'. What happened to problem 'c', Darryl?

"There's no decimal point on the top row and no numbers above '0' and '5', so I couldn't do it," said Darryl.

1. Draw a straight line each column before you solve these problems.

2. Wherever you see a blank space, put a zero in there, so you would have the same number of digits.

"What's eight dollars and five cents?" $ 8|
                                     $ |.0 5
                                   $ 8 .05

"Will you come to the board and do problem 'e' Keesha?"   0. 5 2 3
                                                      3. 1 8 0
                                                      0. 0 1 2
                                                  + 2. 2 0 0
                                                      5. 9 1 5

"Where do you put the decimal point on a whole number like 12?"

"Where I put it in the problem," said Keesha.

Matt raised his hand and said, "After the last digit, Keesha! Remember eight dollars—$8?"

"Oh yeah, thanks," said Keesha.

"Where do you put the decimal point after whole numbers like 8, 20, 115? After the last digit!" said Charles.

"Please correct the problem, Keesha."

```
        0. 5 2 3
        3. 1 8 0
      1 2. 0 0 0
  +     2. 2 0 0
      1 7. 9 0 3
```

(The class clapped for Keesha. She turned around and bowed with a smile.)

"Please look at your Place Value chart carefully. To avoid alignment errors, add '0's so that all numbers in a problem have the same number of digits to the right of the decimal point. Also, write the horizontal problem in vertical format. This will ensure that the place-value positions are aligned correctly."

Skill Practice (noting specific problems)

Examples in addition of decimals.

```
    .5
    .6        Sum of addends is greater than 9.
    .9        Placement of decimal point in the proper
  + .3        place.
    2 .3
```

```
   . 9        Add decimals involving blank spaces.
  +5 .2 8
```

```
   8 .
    .5 3
   2 .1
  +   .6 8
```

.8 2 4 + . 1 9 + 4. 7 5 + 2 0 0   Copy to vertical format; align decimals; placement of decimal in whole number.

"If any of you need to take your folders home in order to study for tomor-row's test, you may do so."

"My mom likes to look over my folder," said Jimmy. "She' s happy that my work is getting better and that I do my homework."

"Yeah, we didn't have folders before and we didn't keep track of our grades or behavior points. Man, we have to watch out for everything in this class!" said Michael.

"Oh yeah, we like it!"

SUMMARY

LIST OF COMMON ERRORS:

1. Arranging numbers in order of their value.

2. Misspelling—hundreds for hundredths; thousands for thousandths.

3. Reducing fractions to decimals, or decimals to fractions.

$$\frac{3}{4} \quad = \quad 0.75 \qquad 0.20 \quad = \quad \frac{1}{5}$$

4. Incorrect placement of decimal point of whole number.

5. Difficulty in adding common fractions and decimals.

6. Ability to add:

$$\begin{array}{r} 7 \frac{34}{100} \\ + 1 \frac{8}{100} \end{array} \qquad \begin{array}{r} 7.34 \\ + 1.08 \end{array}$$

7. Ability to line up the decimals.

8. Ability to place a decimal point after a whole number.

CHAPTER 6

OBJECTIVE 6: SUBTRACTION INVOLVING DECOMPOSITION OF DECIMAL NUMBERS

This objective requires students to solve multi-digit subtraction involving decomposition of digits of higher value and aligning the decimal point.

TEST ITEMS:

a)
```
    4 3 0. 1 4
  -   6 2. 6 8
```

b)
```
    9. 3 5 0
  -   . 2 8 6
```

c)   5 2 3. 9 7 - 2. 6 5

d) $ 1 0 6. 2 9 - $ 9 7. 5 3

e)   9. 0 - 0 . 2 6

ANALYSIS OF ERRORS:

1. Failed to align the decimal points.
2. Failed to regroup.
3. Error due to zero in minuend.
4. Omission of decimal point.
5. Misplacing of decimal point.
6. Misplacing of decimal number in subtrahend.

7. Blank places in minuend—13.3-8.11.

8. Zero difficulties.

The following problems were written on the board to solve:

a)   1 7. 2
   -     1. 6 2 5

b)   .5
   - .3 8 6

c)   .8
   - .0 0 5

d)   2. 7 - 0. 2 7

e)   0. 7 - 0. 5 2 4

Different answers were obtained from the class.
"Deidre, will you do the first problem on the board, please?"

    1 7. 2 0 0
 -      1. 6 2 5
    1 5. 5 7 5

"I added two zeros on the top number because the bottom number had two more; otherwise, I can't subtract," said Deidre.

"Oh, that's what you're suppose to do. Wait a minute, wait a minute, don't do the rest of the problems. Let me redo mine," said Latania. "I thought I learned to add the zeros before when we were doing the subtractions. Shucks, I forgot it this time…that's why I got 1 5. 6 2 5."

"Students, please draw a straight line down each column before you solve the problem. Wherever you see a blank space, put a zero in there. If either the minuend or subtrahend is a whole number, name the number as a decimal numeral. Be sure that the decimal points are placed directly under each other."

"What did you say? Draw a straight line and what?" asked Marvin.

(Marvin has difficulty assimilating more than one direction at a time. He cannot handle an overload of information. On the other hand, Michael would get started quickly and follow the first direction, but fail to listen to the rest of my directions. Michael appears to have short-term memory and is aware of it. I have found that writing on the chalkboard as well as verbalizing and providing a handout sheet has helped these students to keep up with the class.)

1. Draw a straight line down each column before you solve these problems.

2. Wherever you see a blank space, put a zero in there.

(I rewrote the problem with large numbers, drew a straight line down each column, explaining each step out loud—as I went along and added zeros in the blank spaces. The students followed each step, copying it on to their papers.)

$$. 8 \mid 0 \mid 0$$
$$- \quad . 0 \mid 0 \mid 5$$
$$\overline{\qquad \mid \qquad \mid \qquad}$$

"If either the minuend or subtrahend is a whole number, name the whole number as a decimal numeral. Okay, write a whole number on your paper; now make that number a decimal numeral. Where did you put the decimal point?

Good, after the whole number. Now subtract 1.43 from it."

$$1 \; ⑤. 0 \; 0$$
$$- \quad 1. 4 \; 3$$

(The students filled the vacant spaces with zeros.)

"I can't do subtraction this way: 2. 7—0. 2 7," said Darryl. "That's why I left those problems out on the test."

(Some of the students could not keep track of the place value when a problem was written horizontally—a fractional part expressed with a decimal point added to their difficulty.)

"Remember to change the problems that are written horizontally to a vertical format, like you did in addition."

"Yeah, but I forget which number to put down on top. Like in the book it said: 'Subtract 84.317 from 237.2' and my answer was wrong. I put 84.317 on top like I did 2.7—0.27," said Darryl.

"Let's read the problem again: 'Subtract 84.317 from 237.2' (I underlined 'from'). How many do you have?"

"237.2, " responded Darryl.

"Now subtract 84.317. This problem would be written: 2 3 7. 2—8 4. 3 1 7."

On the chalkboard:

$$2\ 3\ 7.2\ \underline{0\ 0} \quad \text{(add 2 zeros)}$$
$$-\quad\ \ 8\ 4.3\ 1\ 7$$

"First, you need to align the decimal points.

Second, you need to add zeros after the 2 in the minuend. This does not change the value of 237.2.

Third, you need to have the same number of digits in the top and bottom row of numbers.

Please review the Place Value Chart you have in your folder. Then, you will understand why you line up the numbers to the right (whole number part) or to the left (fractional part) of the decimal point (and)."

Skill Practice (noting specific problem)

9.60	7.80	6.50	3.05	5.05	9.03
- 7.15	- 4 66	- 3.43	- 1.83	- 3.74	- 5,61

9.00	8.00	7.00	9.64	8.78	6.96
- 2.78	- 4.68	- 3.59	- 6.25	- 4.29	- 3.58

.732	.835	12.35	44.67	22.46
-.489	-.267	- 7.47	- 6.79	- 8.67

"Please check your work, as you did with previous subtraction problems. Since you all understand addition and subtraction of decimal numbers, may we have a test on this tomorrow?"

"Okay, let's take a vote," said Jamal. "Everyone agree? Raise our hands."

"All right, Mrs. Yamate."

"Don't forget our stars. Today is Friday," said Jimmy.

(I have been giving out red, green, and silver stars to students who do not get a detention during the week. The different colors do not have different values. I place the star on the back of each of the student's hand and give a compliment, for example: 'A gold star for Douglas who got an 'A' on his test; 'A gold star for Mark who did all his homework this week', etc. (A tangible reward for specific behavior.) The students have been making designs, lining up the stars, or writing their names with the stars, inside of their folders. They are always busy counting how many stars each of them have and comparing the numbers with their friends.

I started passing out stars as a means of classroom management. My students were labeled 'behavior problems', 'low achieving', and 'learning difficulties'; therefore, I started pulling the classroom together from the first day of class by letting them know that I am the teacher; I am here to help them learn mathematics and no learning takes place in a disorderly classroom—be it a physical environment or emotional/mental state of students. The students need to be 'marching' forward with me. This is a remedial math class—low track, and they may view me as their 'hope' for understanding and learning math, since I tech the higher level classes, too. For these students to gain their self-respect, self-confidence, and esteem among their peers is critical to their being as adolescents, disadvantaged learners. Additionally, I advised them that their parents will be coming to "Back to School Night", and I will have the folders available for them to inspect their behavior and learning progress.

My intention was to slowly diminish the reward from every week to every other week, then periodically, and finally extinguish this type of reward system. However, the students looked forward to earning their stars and would remind me if I did not pass them out; I continued to pass them out all the way to the end of school. Result: Classroom management has been rewarding for me as well as for the students—minimal problems.)

The following week

"May we sit in a circle today?" said Stephen.

"No, this is test day," said Matt.

"You won't trust us, huh? Do you think we might cheat?" asked Stephen, not waiting for my reply.

"It's all right with me, if you all want to sit in a circle. Cheating off of your neighbor's paper only means cheating yourself. Your neighbor may have all the information in his (or her) head and you may not. Are you going to drag your friend around with you for the rest of your life because you want to use his (or her) brain?"

"Hey, Stephen, shall we become Siamese twins?" said Matt.

(Stephen and Matt stood up in front of the class, tilted their heads as if attached as Siamese twins. The class laughed. The students come up with clever acts and 'raps'—treat the students with respect and humor, then they respond with open hearts!)

SUMMARY

ANALYSIS OF ERRORS

1 6. 2 - 4. 6	Regrouping tens.
5. 6 1 - . 9	Blank space in the subtrahend.
5. - 1. 4 3	Blank space in the minuend.
6. 5 3 1 - 5. 9 7 5	Regrouping all places in minuend.
9 - .0 5	Renaming a whole number as a decimal number.

CHAPTER 7

OBJECTIVE 7: MULTIPLICATION OF DECIMALS

This objective requires student to place the decimals in the proper place in the product and prefix or annex zeros when necessary.

TEST ITEMS

a)
```
    7 8. 8
x     4 2
```

b)
```
  0. 3 2 2
x       5 3
```

c)
```
  0. 0 0 7
x     0. 8 7
```

d) 0. 4 6 x 0. 6

e) 0. 6 x 0. 0 7

LIST OF COMMON ERRORS:

1. Misplacing of decimal point.

2. Omission of decimal point.

3. Failure to annex zero.

4. Annexing of unnecessary zero in multiplier.

5. Aligning decimal points (multiplier and multiplicand).

6. Error in multiplication.

EFFECTIVE STRATEGY—Mastery of Multiplication of Decimals (while carrying on dialogue with students).

The following problems were written on the board to solve as a class:

a)   4 6   b)   .2 3 8 9
     x .0 4   x   2

c)   .0 2   d)   1. 3 5
     x .0 3   x   . 0 6

e)   2. 6 x .0 0 2   f)   .0 7 x $\frac{1}{2}$

"Class, let's check our answers if you are all finished."

The students started to answer: "1. 84; 0.4778; 0.0006; .06…someone is wrong!" said Eric

"Let's do 'c' together so we can see why we differ in our answers."

    .0 2   (2 decimal digits)
  x .0 3   + (2 decimal digits)
    0 6   (4 decimal digits)
    0 0
  .0 0 0 6   (count over 4 decimal digits from the right to the left)
  ← ← ← ←

"Aren't you supposed to bring the decimal straight down? That's what we did in addition of decimals?" asked Marvin.

"In multiplication of decimal digits, you will find it easier to follow the steps:

1. Do the multiplication of the 2 decimal numbers.

2. Count the number of decimal digits in the multiplicand and multiplier (2 places to the right of decimal in both multiplicand and multiplier).

3. Add the number of decimal digits—2 decimal digits + 2 decimal digits = 4 decimal digits.

4. sum of the decimal digits equals the number of decimal digits in the product.

"Marvin, we are multiplying $\underline{\quad 2 \quad}$ x $\underline{\quad 3 \quad}$ = $\underline{\quad 6 \quad}$ "
$$ 100 100 10000 \, .$$

"Okay, I get it now."

"Class, let's first count the number of places to the right of the decimal in the multiplicand (top number) and the number of places to the right of the decimal in the multiplier (bottom number). The *sum* of the decimal digits in the multiplicand and multiplier will give you the number of decimal digits in the product—move from right to the left. Count out loud with me, 1, 2, 3, wait a minute, we don't have any more numbers. We have to *prefix* or *annex* a *zero* in order to have a 4th place. Now, place the decimal point.

$$
\begin{array}{r}
1.\,3\,5 \\
\text{x} \quad .\,0\,6 \\
\hline
8\ 1\ 0 \\
0\ \ 0\ 0 \\
\hline
.0\ \ 8\ 1\ 0 \\
\leftarrow \leftarrow \leftarrow \leftarrow
\end{array}
$$

(2 decimal digits)
+ (2 decimal digits)
(4 decimal digits)

(count over 4 decimal digits from right to left)

You don't have to place the decimal point in the partial product. We can use a shortcut method for this problem, so you can save time in a test.

$$
\begin{array}{r}
1.\,3\ 5 \\
\text{x} \quad .\ 0\ 6 \\
\hline
.0\ 8\ 1\ 0
\end{array}
$$

Please copy this example down in your notes so you can understand the placement of the decimal point. I placed the decimal point in the partial product just to show you what you are multiplying:

```
        8. 3 3
    x   1  2. 6
        4. 9  9 8    ( .6  x  8. 3 3)
    1 6. 6  6        ( 2  x  8. 3 3)
    8 3. 3           (10  x  8. 3 3)
  1 0 4. 9  5 8
```

"What did you get for problem 'f'?"   Problem f: .07 x $\frac{1}{2}$

"Mrs. Yamate, you can't mix a fraction with a decimal," said Lhyn.
"What's half of a dollar?"
"Fifty cents," said Lhyn.
"All right. Change the $\frac{1}{2}$ to .50 and multiply. What's quarter or one-fourth of a dollar?"
"Twenty-five cents," said Lhyn.

Skill Practice (noting specific problem)

Changing fractions to decimals.

$\frac{3}{4}$ =   .75 $\frac{1}{3}$ =   .33 $\frac{1}{3}$

$\frac{1}{2}$ =   .50 $\frac{2}{3}$ =   .66 $\frac{2}{3}$

$\frac{1}{4}$ =   .25

$\frac{1}{5}$ =   .20

"Remember when I taught you that the fraction bar can be read as 'divided by'—as 1 divided by 2?"
"You can't divided 1 by 2 because it's a small number," said Lhyn.

"All right, follow me. Let's divide: $\underline{1}$ $\underline{0.5}$

2     $2\overline{\smash{)}\,1.0}$

Put a decimal point after the whole number and add one zero in the dividend. If you write the denominator (divisor) as I wrote it, you will not get mixed up—as to which number goes into the box."

"Now Andrew, will you please come to the board and the last problem?"

$$.07 \text{ x } 25 = \quad 1.75$$
$$\underline{\text{x} \quad .5} \text{ (½)}$$
$$.875$$

"It's easier for me to multiply the decimals vertically," said Andrew.

"Hey, Andrew, let me come up and help you. You kind of got mixed up in writing the problem," said Stephen.

"It's okay. I'll go and help him. I was his partner," said Samuel.

Samuel's correction:

$$.07 \text{ x } .50 = .0350 \qquad\qquad\qquad .50 \text{ (1/2)}$$
$$\underline{\text{x} \quad .07}$$
$$.0350$$

<u>Skill Practice (noting specific problems)</u>.

Examples in multiplication of decimals.

SUMMARY

LIST OF ERRORS

a)   4 x .2   Change a whole number to a decimal number.

b)   1 4   Prefixing zeros.

$\underline{\text{x} .007}$

c)   .5 8   Carrying in ones place. Counting number of
     x   .6   places after the decimal for right placement
                                  in product.

d)   .0 1 1 x .0 5

"If anyone needs additional help, please come to the chalkboard. I will write some problems for you to solve. Then, I can correct any misunderstanding immediately."

(Four students came up to do the problems on the chalkboard. This enables me to observe the student's application of the procedure to the problem and correct his/her thinking process immediately.)

CHAPTER 8

SET UP 8: DIVISION OF DECIMALS

This objective requires students to solve long division with decimals in the divisor and in the dividend, name the quotient, do multiplication of the quotient involving carrying, and do the necessary subtraction.

TEST PROBLEMS

a) 7 | 7 1. 5 4

b) 0. 6 | 0. 5 0 4

c) 3. 2 | 4 9 5 3. 6

d) 6 9 2. 5 ÷ 2. 3

e) $2 6. 1 0 ÷ 1 5

LIST OF COMMON ERRORS

1. Misplacing of decimal point.

2. Omission of decimal point.

3. Decimal point used when unnecessary.

4. Failure to annex or prefix zero in quotient.

5. Failure to annex zero to dividend.

6. Errors in division, multiplication, and subtraction.

8. Moving decimal point only in divisor.

EFFECTIVE STRATEGY—Mastery of Division of Decimals (while carrying on dialogue with students).

"We are ready to move on to division of decimals today. Will you please work on the worksheet that you picked up from the front desk when you entered?"

"May we sit in a group today, Mrs. Yamate, since we're working on a worksheet?" said Kenya.

"Yes, you may."

"Elton, why are you, Lhyn, and Antarra talking?"

"Well, this is the truth. We started off by talking about our math problem, you know, then I looked in Lhyn's mouth and noticed her braces. Then, we got talking about braces on your teeth."

"How did the subject get changed from math to braces on your teeth?

"Well, I asked her how it feels. Then, Antarra joined in. You see, I'm going to have braces put on my teeth, so I wanted to know if it hurts."

"Well, what did Lhyn say?"

"She said they hurt when the doctor put them on and for couple of days after, but she got used to them," said Elton.

"You can go on with your work now, right?

"Yes."

Examples of thirty problems on worksheet:

$$7 \,|\, \overline{8\;\;6.\;\;1}$$

$$6 \,|\, \overline{6\;\;7\;\;6\;\;5.\;\;6}$$

$$2 \,|\, \overline{.8\;\;6\;\;0\;\;8}$$

"Most of you got all the problems correct; you knew that you place a decimal point in the quotient directly above the decimal point in the dividend. Some of you did not put the decimal point in the quotient. Did you forget?"

"Yeah," said Sean.

"I see that you are all doing short division problems the long way. Let me show you how you can save time on a test doing one digit divisor problems.

You use a crutch like you do in carrying:

$$7\ |\ 8 \quad {}^1 6. \quad {}^2 1$$

The following problems were written on the blackboard:

$$6\ |\ .0\ \ 3\ \ 6 \qquad\qquad \text{Prefix zeros in quotient.}$$

$$.2\ |\ 6 \qquad\qquad \text{Annex zero to dividend.}$$

$$.2\ |\ 6.\ \ 0 \qquad\qquad \text{Annex zero to quotient.}$$

"Ready to check our answers, class?"
".6; .06," the class responded.
"You forgot to prefix with *two zeros* in the quotient.

$$\frac{.0\ \ 0\ \ 6}{6\ |\ .0\ \ 3\ \ 6}$$

Be sure and place a number in the quotient directly above each digit in the dividend. Please multiply to check your answers—multiply the quotient by the divisor."

$$\begin{array}{r} .0\ \ 0\ \ 6 \\ \times \qquad\quad 6 \\ \hline .0\ \ 3\ \ 6 \end{array}$$

"What did you get for the next problem?"

"3; 3.0" the class responded. $\dfrac{3}{.2\ |\ 6}$

"How did you get 3…where did you place the decimal point?"
"There's no decimal point inside the box," said Darryl.
"Please write the following procedure onto your notes (on the chalkboard):

1. If the divisor has a decimal point, make the divisor a whole number by moving its decimal point to the right. Place a decimal point after the whole number in the dividend and annex a zero. Now, move the decimal point one place to the right. You may use a carat to indicate your new position.

$$
\begin{array}{r}
3 \quad 0. \\
.2 \mid \overline{6. \quad 0} \\
 {}^{\wedge} \qquad {}^{\wedge}
\end{array}
$$

2. Move the decimal point in the dividend to the right as many places as you moved the decimal point in the divisor. You may use a caret (^) to indicate your new position.

3. Divide as in the division of whole numbers and place the decimal point in the quotient directly above the caret in the dividend.

(I read the whole procedure aloud, and indicated as I proceeded doing the example, as moving the decimal point to the right. Then, I had the students say out loud the procedure for moving the decimal point, as they did the following example:

$$
\begin{array}{r}
\overline{\qquad\qquad .\_} \\
.4 \mid 3 \ 5. \ 6 \\
 {}^{\wedge} \qquad\qquad {}^{\wedge}
\end{array}
$$

"Move the decimal in the divisor one place, put a caret there. Move the decimal in the dividend one place, put a caret there. Put a decimal point in the quotient directly above the caret in the dividend. Now, divide," said each student.

(I find that the students learn and remember better if each of them would write the problem, do it, and verbalize during the course of solving the problem. After they learn the procedure, they do not verbalize.)

"Students, we've been moving the decimal points because we want to work with a whole number in the divisor. What are we doing when we move the decimal point one place in the divisor, for example, making .4 to become 4? Why is this possible?"

"Just move it!" said Sean and Andrew.

"You cannot just pick up the decimal point and say, "We don't want the decimal point in the divisor, so we're going to pick it up and move it."

"Well, that's what we did before. Why is there a reason why we can't?" asked Antarra.

"Will you all please multiply 0.4 x 10. What is the answer?"

"We got 4," said Charles and Jamal.

"If we multiply the divisor by 10, then we have to multiply the dividend by 10.

Now the problem becomes:
$$4 \overline{\smash{)}3\ 5\ 6}$$

Remember, we are not just picking up the decimal point and moving it. What are we multiplying by when we move the decimal point two places?"

"You're multiplying by 100," said Charles and Jamal.

"Correct. Our homework will have similar problems. You may start your work now so I can go around and check your work, to make sure you're on the right track. Please pass the worksheets out now, Mark."

"Come here, Mrs.Yamate," said Eric

While approaching him I said, "Why don't you come to see me?"

"Well, I don't want you to get arthritis," said Eric. Eric is 'chubby' and fits snugly into his chair.

Andrew moved his desk next to mine. "Watch me as I do this, Mrs. Yamate. I never could learn it before. I want to make sure that I am doing it right."

(Gregg was rotating his head while doing his work.)

"Gregg, did you lose a screw off of your neck?"

"No, sorry," he said and smiled.

Skill Practice (noting specific problems).

$$5 \overline{\smash{)}.0\ 1\ 5} \qquad \text{Prefix zeros.}$$

$$2\ 0\ 0 \overline{\smash{)}4} \qquad \text{Change to decimal numeral and add zeros.}$$

$$.6\ 2\ 5 \overline{\smash{)}1\ 5} \qquad \text{Multiply divisor and dividend by thousand. (Move decimal 3 places.)}$$

$$4\ |\ \overline{1.\ 2\ \ 8\ \ 4}$$ No prefix zero.

$$.9\ 6\ \ |\ \overline{8.\ \ 8}$$ Annex zero in dividend.

(Before I could write "Conference Sign Up" on the chalkboard, the students had written it and their names underneath. Antarra, Michael, and Sean brought their desks to the front, next to mine.)

"What's the matter, Antarra, Michael, Sean?"

"I could never do division when there were more than one number in the divisor, so I want to sit close to you," said Antarra.

"I don't want to get a detention. I could study better when I sit near you," said Michael. "I've been in enough trouble in school today."

(Antarra and Michael are taking pride in their accomplishment. They would have preferred to sit way in the back of the classroom before, but now, they want to sit next to the teacher!)

"We'll spend a little more time on division of decimals before we have a test. We all need more practice. Why don't we have a clean up day today, class?"

"What do you mean by a clean day" asked Jennifer.

"I mean clean up questions you may have in your heads, or errors that you made in your homework, etc."

"Yeah, she's going to sweep it out of our heads," said Kenya.

"Everyone, check your folders for any papers that you haven't corrected. We are going to have a test on division of decimals tomorrow. I think I will add some multiplication of decimals for review, too. Okay, as I call your names please bring your folders up with you and let me review with each of you the materials in your folders."

(I check the folders with each student every other week. They take their folders home every Monday. Why on Monday? The students will not forget to bring them back the next day, and the parent/guardian would have time to look over the work. If I send it home on Friday, they may not come back on Monday)

C H A P T E R   9

SET UP 9: ADDITION OPERATION WITH FRACTIONS

This objective requires students to add fractions with unlike denominators and express the answers in simplest terms.

TEST PROBLEMS

a) 6 $\dfrac{2}{3}$

    4 $\dfrac{1}{6}$

 +

b) 2 $\dfrac{1}{5}$

    2 $\dfrac{1}{10}$

 +

c) $\dfrac{11}{25}$ + $\dfrac{3}{50}$ + $\dfrac{37}{100}$

d) 8 $\dfrac{1}{3}$

      $\dfrac{4}{7}$

 +

e) 5 $\dfrac{1}{2}$ + 2 $\dfrac{1}{2}$

LIST OF COMMON ERRORS:

1. Added the denominators to get the denominator for the sum.

2. Forgot to add the whole numbers in adding mixed numbers.

3. Error in common denominator—wrong Least Common Denominator.

4. Forgot to simplify.

EFFECTIVE STRATEGY—Mastering of Like Denominators (while carrying on dialogue with students).

"We are going to shift from whole numbers operation to fractions. Let's do the following problems on the chalkboard together."

$$
\text{a)} \quad \begin{array}{r} \frac{1}{3} \\ + \quad \frac{1}{3} \end{array}
\qquad
\text{b)} \quad \begin{array}{r} \frac{1}{4} \\ + \quad \frac{1}{4} \end{array}
$$

$$
\text{c)} \quad \begin{array}{r} \frac{3}{4} \\ + \quad \frac{1}{4} \end{array}
\qquad
\text{d)} \quad \begin{array}{r} \frac{3}{5} \\ + \quad \frac{4}{5} \end{array}
$$

$$
\text{e)} \quad \begin{array}{r} \frac{4}{5} \\ + \quad \frac{4}{5} \end{array}
\qquad
\text{e)} \quad \begin{array}{r} \frac{5}{6} \\ + \quad \frac{5}{6} \end{array}
$$

"May I have the answer for the first problem, class?"

"$\frac{2}{6}$; $\frac{2}{3}$; $\frac{1}{3}$"

"Keesha, will you please show us how you got $\frac{2}{6}$?"

"I added the 1s and drew a line and added the 3s," said Keesha. "It's addition!"
"Yeah, and I reduced the $\frac{2}{6}$ to $\frac{1}{3}$, said Marsha."

Problem shown on the board: 1   third
                           + 1   third
                              2   third

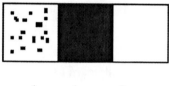

$$\frac{1}{3} \; + \; \frac{1}{3} \; = \; \frac{2}{3}$$

"If the denominators are the same, you just add the numerators and write the sum over the common denominator. Then, change the answer to simplest form. I see many of you are taking notes today. Very good!"

"What did you get for 'b'?"

"$\frac{2}{4} = \frac{1}{2}$", responded the class.

"What did you get for 'c'?"

"$\frac{4}{4}$ No, 1," responded two different groups.

"Whenever the numerator and the denominator are the same it is equal to '1', so $\frac{7}{7} = 1, \frac{5}{5} = 1$ (reduces to a natural number, 1). Remember, the bar between

the numerator and the denominator (7) is also a symbol for division."

"What do you call a fraction like $\frac{1}{4}$?"

"Don't know," responded the class.

"We didn't learn all this stuff before, Mrs. Yamate, but if you teach us, we'll learn it! Right, class?" commented Jamal.

We want to get into them algebra books," said Charles.

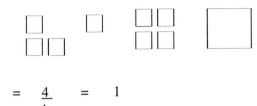

$$\frac{3}{4} \;+\; \frac{1}{4} \;=\; \frac{4}{4} \;=\; 1$$

"When the <u>numerator</u>   (which gives the number of parts)
          denominator (gives the kind of part or how the whole was
                        divided)
is smaller than the denominator, it is called a *proper* fraction. Please add these words to your vocabulary list. All right, what did you get for 'd'?"

"$\frac{7}{5}$; 1 $\frac{2}{5}$."   (Two different answers.)

"How many $\frac{5}{5}$ are in $\frac{7}{5}$? "

"1," chorused the class.
"How many fifths are left?"
"Two-fifths," answered the class.
"You can reduce the *improper* fraction to a *mixed* number, 1 $\frac{2}{5}$. Try dividing 7 by 5. You call it a mixed fraction because you have a whole number and a fraction. Now, check this out. Do the same thing with your construction paper so you can understand fractions like I did:

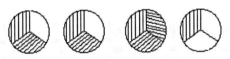

"Did you get 1 $\frac{3}{5}$ for the next answer?"
"Yes. We know how to do these, now," said Andrew.
"I got 1 $\frac{4}{6}$ for 'f', said Jason.
"No, it's 1 $\frac{2}{3}$. You forgot to reduce the $\frac{4}{6}$" said Charles.

"There are as many as five different kinds of answers in the addition of fractions. These are:

1. Proper fractions not reducible to lower terms.

2. Proper fractions that are reducible.

3. Improper fractions reducible to natural numbers.

4. Improper fractions reducible to lower terms, which can be changed to mixed numbers.

5. Improper fractions, not reducible, which can be changed to mixed numbers.

"May I have an example of number 1?"
"$\underline{2}$"
3, chorused the class.

"May I have an example of number 2?"
"$\underline{4} = \underline{1}$"
8   2, chorused the class.

"May I have an example of number 3?"
"$\underline{5} = 1$," chorused the class.
5

"May I have an example of number 4?"
"$\underline{10} = \underline{5} = \underline{2}$"
6   3   1 3,   chorused the class.

"May I have an example of number 5?"
"$\underline{7} = \underline{1}$"
6   1 **6,**   chorused the class.

"We will now work on the practice worksheet which has similar problems:

$$\begin{array}{ccccc}
\frac{2}{5} & \frac{5}{8} & \frac{13}{32} & \frac{9}{16} & \frac{17}{24} \\
\frac{2}{+\,5} & \frac{5}{+\,8} & \frac{15}{+\,32} & \frac{10}{+\,16} & \frac{19}{+\,24}
\end{array}$$

"You may work with a partner or work in a group."

"Mrs. Yamate, I missed a lot of school when I was in the elementary school, so I never did learn this," said Kenya. "I don't know anything about fractions. I need some more help."

"Students, I am going to have a group come up to the front chalkboard. Anyone who needs more practice come to the front."

(One problem was written for each student on the chalkboard to solve. There were ten of them. As each finished his/her problem, I corrected them immediately.)

"Your homework is going to be the next level from what you are doing. Let me go over some problems with you now."

$$6\frac{3}{5} \qquad 8 \qquad 7\frac{2}{9}$$
$$+\ 7 \qquad\qquad +\frac{7}{8} \qquad 2\frac{5}{9}$$
$$13\frac{3}{5} \qquad 8\frac{7}{8} \qquad +\frac{9}{9}$$
$$\qquad\qquad\qquad\qquad 9\frac{7}{9}$$

$$8\frac{8}{8}$$
$$\frac{7}{8}$$
$$+\quad$$
$$8\frac{15}{8} = 8 + 1\frac{7}{8} = 9\frac{7}{8}$$

"I thought you are supposed to put an $\underline{8}$ in the second problem because there
8
is no fraction on top," said Patrick.

"You are thinking of subtraction of fraction when you decompose or rename the 8 to 7 $\underline{8}$ We'll get to that when we do subtractions."
$\quad 8\ .$

"Class, the chart that Douglas is passing out will be helpful for you to make a list of as many groups of equivalent fractions as you can discover. This will make it easy for you to change unlike fractions to like fractions when subtracting or adding them. Don't lose it!

"We will go on and add unlike denominators since all of you seem to understand addition of like denominators. Wow, it makes me feel so good when you all do well. Thank you, class. This is a teacher's reward."

(The students sat up straight in their seats and looked proud.)

EFFECTIVE STRATEGY—Mastering of Unlike Denominators (while carrying on dialogue with students).

"We are going to shift to adding unlike denominators today.

$$\frac{3}{4}$$
$$\frac{5}{8}$$
$$+\ \frac{1}{2}$$

In this example, the common denominator is included among the given denominators. Eight is a multiple of '2' and '4', therefore, the common denominator is '8'. Let's change the fractions to equivalent fractions having a common denominator:

$$\frac{3}{4} = \frac{3}{4} \times \frac{2}{2} = \frac{6}{8}$$

$$\frac{5}{8} = \frac{5}{8}$$

$$+\ \frac{1}{2} = \frac{1}{2} \times \frac{4}{4} = \frac{4}{8}$$

Now that we have changed the fractions to equivalent fractions, we can add. When you multiply the numerator and the denominator of a fraction by the same number as $\frac{2}{2}$ and $\frac{4}{4}$ it is equivalent to multiplying that fraction by

'1'. Any number multiplied by '1' is the number itself.

"Remember in the test there was a problem:
$$\frac{11}{25}$$
$$\frac{3}{50}$$
$$+\ \frac{37}{100}$$

You all can do it now, right?"

"Yeah. Let us try it now," said Mark.

"What did you get?"

"How did you get $\frac{87}{100}$?" said Darryl.

"Look, Darryl. Twenty-five goes into 100 four times, right? So you multiply the numerator 11 and the denominator 25 by 4. Fifty goes into 100 two times, so multiply the numerator 3 by 2 and the denominator 50 by 2. Now all the denominators are 100, so you can add," said Mark.

"Mark, the word 'goes into' bothers me. Twenty-five does not 'go into' 100. How many sets of 25's are there in 100?"

"See if you can do the following problem like you did the last:

$$3\frac{5}{12}$$
$$= \quad 3\frac{5}{12}$$
$$+ \quad 4\frac{1}{4} \quad = \quad 4\frac{1}{4} \quad x \quad \frac{3}{3} \quad = \quad 4\frac{3}{12}$$

I see that some of you forgot to simplify, reduce. You can use prime numbers as a test for reducing.

Remember: a) all even numbers are divisible by 2.

   b) a number is divisible by 4 if the number formed by its last two digits is divisible by 4.

   c) a number is divisible by 3 if the sum of its digits is divisible by 3.

   d) a number is divisible by 8 if the number formed by the last three digits of a number is divisible by 8.

   e) a number is divisible by 5 if it ends in 5 or 0.

   f) a number is divisible by 9 if the sum of its digits is divisible by 9.

   g) a number is divisible by 11 if the sum of its digits in the *odd* places, counting from the right, minus the sum of its digits in the *even* places, is divisible by 11.

"This Test for Divisibility should be in your folders. Please check. If you do not have it, please copy it down again and keep it as a guide when you reduce. If you practice doing this over and over again, simplifying large numbers will become automatic and fun!"

The following problems were written on the board for the class to solve together:

Add: a) $\dfrac{1}{2} + \dfrac{1}{3}$

b) $\dfrac{1}{2} + \dfrac{2}{3} + \dfrac{3}{5}$

c) $\dfrac{1}{3} + \dfrac{1}{4} + \dfrac{1}{6}$

"The common denominator for 'a' is 6, the product of 2 and 3. We got $\dfrac{5}{6}$ for our answer," said Sean and Lhyn. "Did any other group get a different answer?"

"No, that's an easy one," responded the class.
"What do you do when you have three different denominators?" asked Mark.
"You can see that 2 and 3 have 6 in common but 5 doesn't, so multiply the three numbers," said Lhyn. "I'll write on the chalkboard for you:

$$\dfrac{1}{2} + \dfrac{2}{3} + \dfrac{3}{5} = \dfrac{15}{30} + \dfrac{20}{30} + \dfrac{18}{30} + \dfrac{53}{30} = 1\dfrac{23}{30}$$

"What is the denominator for 'c'?"
"The common denominator must contain 3, 4, and 6 as divisors. Since 6 is contained in 3 x 4, 12 is the common denominator. We got $\dfrac{9}{12} = \dfrac{3}{4}$," said Marsha.

"Class, to find the lowest common denominator (LCD) of two or more fractions, list the multiples of the denominators until a multiple common to both numbers is found. That multiple is the LCD of the two fractions.
For example: $\dfrac{1}{3}$ and $\dfrac{4}{7}$

Multiples of 7 are: 7, 14, *21*
Multiples of 3 are: 3, 6, 9, 12, 15, 18, *21*
Therefore, 21 is the LCD. If this is confusing, do this:

Multiples of 7:   7, 14, 21, ... can I divide 7 by 3?   No

can I divide 14 by 3? No

can I divide 21 by 3? Yes

Therefore, 21 is the LCD."

"Assignment for tomorrow is addition of like denominators and unlike denominators I will go over the problems again with anyone who needs additional help at the chalkboard, so please come up. The rest of you may do the worksheets now. Would you agree that Friday will be our test day? Anyone needing additional help, please see me now."

CHAPTER 10

SET UP 10: SUBTRACTION OPERATION WITH FRACTIONS

This objective requires students to subtract fractions with unlike denominators and express the answers in simplest term.

TEST PROBLEMS

a) $8\dfrac{7}{8}$
   $-\ 3\dfrac{1}{4}$

b) $27 - 26\dfrac{1}{6}$

c) 6
   $-\ \dfrac{1}{5}$

d) $5\dfrac{3}{4} - 5\dfrac{1}{8}$

e) $\dfrac{7}{8} - \dfrac{3}{8}$

LIST OF COMMON ERRORS:

1. Error in regrouping.

2. Prefixed the decomposed lower unit number to the numerator of the fraction.

3. Added the decomposed lower unit number to the numerator of the fraction.

4. Failed to decompose the whole number.

EFFECTIVE STRATEGY—Mastering of Like Denominators (while carrying on dialogue with students).

"We have been putting fractions together until now. We will do just the opposite today—take them apart. Let's do the following problems on the board."

a) $\frac{5}{8}$
    $\frac{1}{8}$
$-$

b) $\frac{5}{8}$
    $\frac{2}{8}$
$-$

c) $\frac{9}{16}$
    $\frac{3}{16}$
$-$

d) $\frac{7}{10}$
    $\frac{3}{10}$
$-$

e) $\frac{5}{6}$
    $\frac{1}{6}$
$-$

f) $\frac{53}{64}$
    $\frac{29}{64}$
$-$

"Marvin, would you like to read off the answers? Everyone check your own answers, please."

"Can I stand up in front of the class—you know, to be a teacher's Assistant?" said Marvin.

(Acknowledging student's need for importance among classmates.)

"Certainly! It's a good idea."

" $\frac{1}{2},$ $\frac{3}{8},$ $\frac{6}{16},$ " said Marvin.

"You forgot to reduce $\frac{6}{16},$ " said Marsha.

"Look, you can't divide 16 by 6," said Marvin.

"You can reduce it by dividing 6 and 16 by 2. Right, Mrs. Yamate?" said Marsha. "They are even numbers!"

"Oh, you're right. I keep forgetting that if the bottom and the top number cannot be divided by the top number, I need to look for another number that both of them can be divided by. You're right," said Marvin nodding his head.

$$\frac{6}{16} \div \frac{6}{6} \quad \text{doesn't work.} \quad \frac{6}{16} \div \frac{2}{2} = \frac{3}{8}$$

"Okay, 'd' is $\frac{2}{5}$, 'e' is $\frac{2}{3}$, and 'f' must be $\frac{3}{8}$ too, since $\frac{24}{64} = \frac{6}{16} = \frac{3}{8}$," said Marvin.

"I got 'em all right!" yelled several students.

"Can we put an 'A' on our papers?" asked Walter.

"What if you got one wrong? Can I write 'A' on my paper?" asked Marvin.

"No, that's an 'A-'," said Walter. "Oh, it's okay, you can write an 'A' Marvin since you read off the answers."

(Peers assessing each other.)

"Aren't we smart, Mrs. Yamate?" said Keesha.

"Yes! Give yourselves a big clap!"

(Students clapped in unison three sets of 3 claps.)

"Since you all know how the to find the common denominators, we will subtract fractions with different denominators now. Let's do the problems on the chalkboard together."

a) $\frac{7}{8}$
$-\frac{1}{4}$

b) $\frac{3}{4}$
$-\frac{7}{12}$

c) $\frac{5}{6}$
$-\frac{1}{3}$

d) $\frac{2}{3}$
$-\frac{3}{5}$

"Remember, you have to look for the common denominator. What's the common denominator for 'a'; and 'b'?

"8 for 'a'; 12 for 'b'," responded the class.

"Find the common denominator first, and then reduce your answer if possible," said Patrick and Andrew.

"Let's check our answers."

"Mrs. Yamate, it's the girl's turn to read off the answers. May I read them?" asked Tyshenna and Deidre.

"Thank you for reminding me, Tyshenna and Deidre. It is the girl's turn. Please come up to the front and read off the answers."

"I have $\underline{5}$ for 'a', $\underline{1}$ for 'b', $\underline{1}$ for 'c', and $\underline{1}$ for 'd'."
    8   6   2   15

"Something's funny, Mrs. Yamate. I didn't get the same answers," said Kenya.

Tyshenna bent over towards Kenya to look at her paper. "Hey, you added instead of subtract," said Tyshenna.

"I guess I got mixed up because I was busy changing the denominators like when we did the addition," said Kenya.

(Kenya seems to have problems shifting from one operation to another. Several of the students would cling on to an algorithm or a rule taught, use it inappropriately for solving problems where it does not apply.)

"Kenya, see if putting a circle around the operation sign first would help you to remember."

Skill Practice (noting specific problem).

$$1 \qquad 3 \qquad 7\,\tfrac{2}{3} \qquad 3$$
$$-\ \tfrac{1}{2} \qquad -\ \tfrac{1}{4} \qquad -\ 3 \qquad -\ 1\,\tfrac{1}{2}$$

"What do you have for the first problem, Fred?"

"1 $\underline{1}$ " said Fred.
   2,

(I demonstrated with a piece of paper and wrote on the chalkboard: '1'. "Now Fred, I'm going to give you $\underline{1}$ of this paper. What shall I do next?"
                                  2

"Tear it up in two," said Fred.

(I wrote on the board: $1 = \underline{2}$) "Now I can give you $\underline{1}$ How many piece do
                           2   2.
I have left? "

$$1 = \frac{2}{2}$$
$$\frac{1}{2} = \frac{1}{2}$$
$$-\ \ \ \ \ \ \ \ \ \frac{1}{2}$$

"Michael, will you please come up and do the next problem?"

$$3$$
$$-\ \frac{1}{4}$$

Michael's work:

$$\frac{2}{3\ 2} = \frac{4}{3\ 4}$$
$$-\ \frac{1}{4} = -\ \frac{1}{4}$$
$$\frac{3}{3\ 4}$$

"Why did you put $\underline{2}$ next to 3?"
            2

"Because that's what you did with the first problem. You need a fraction to take away a fraction," replied Michael.

"I'm sorry, class. I didn't make myself clear. Let's take *three* pieces of paper. I want to give you $\underline{1}$ of a piece. What do I need to do?"
                        4

"Cut up *one* piece of paper into four pieces," said Michael.

"Now, I have 2 whole pieces left and I cut the one paper into 4 pieces so I can give you $\underline{1}$. In the first problem, I had to give you $\underline{1}$ so I had to cut up
            4   2,
the paper into two pieces. The denominator of a fraction determines into how many pieces you need to cut up the whole piece."

$$9 \qquad 7 \qquad 8 \qquad 6 \qquad 12$$
$$-\ \frac{1}{8} \quad -\ \frac{1}{6} \quad -\ \frac{2}{3} \quad -\ \frac{4}{9} \quad -\ \frac{7}{10}$$

Let's do these problems together. I regroup 9 and make it   8 $\frac{8}{8}$

because the denominator in the subtrahend is 8. Now, I can subtract $\frac{1}{8}$

and the answer is 8 $\frac{7}{8}$.

(Rest of the problems were done step by step together out loud.)
"Darryl, will you come and do the next problems?"

$$
\begin{array}{ccc}
7\,\frac{2}{3} & = & 7\,\frac{2}{3} \\
& = & \frac{3}{3} \\
-\;\;3 & & 2\,\frac{3}{3}
\end{array}
$$

"I don't know what to do from here," said Darryl.
"Shucks, I am not doing a good job of explaining today. I apologize."
"Mrs. Yamate, are you sure you know how to do this?" asked Stephen.
"Of course she does!" said the class.
I lifted my shoulders up and smiled. "One of those days."
"Darryl, if you have a board 7 $\frac{2}{3}$ feet long and you cut off 3 feet of it,

how many feet is left of your original board now?"

"I get it now. I have 4 $\frac{2}{3}$ feet left. It's only when you don't have a fraction

in the top number that you break up 1 and make it into a fraction," said
Darryl. "I'll put a cloud around the top number again."

"Everyone, please look at you next problem. Check it over with your partner *and then* give me the answer. I don't want to be a failure today. I'd like to
get the correct answer."
(The class laughed.)
"Oh, it's okay. You always tell us that failure is not final. You always tell us
that *if we fail, we have to take steps to correct our mistake, so that's what we
are doing in your class, right?*" said Jamal.
"Yeah, Mrs. Yamate. It's okay. We get it now," said Darryl and Jimmy.
"What is the solution for the next problem?"

"Hey you guys, did you all get 1 and $\underline{1}$?" asked Samuel.
$$2$$
"Yes. We know how to do this at last! We never learned subtraction of fraction like this before," responded the class.

"Thank you, class. Now, your homework assignment will have the same type of problems. You may get started on them now. I need a rest!"

(The real person comes through!)

"Please sign up for conference if you need further help. I will give you the answer for the first three problems when you finish, so please let me know. I don't want to see you waste your time making mistakes because you didn't understand my explanation. I need to wind up my brain, too."

"Check our papers, Mrs. Yamate," said Samuel and Andrew. "We know how to do these now."

"They are all correct."

"We can be Teaching Assistants for you today," said Samuel and Andrew.

"Great! Thank you. Class, Samuel and Andrew will go around to help you, so raise your hands. They are going to be Teaching Assistants today."

(Response to student's ideas. Peer coaching.)

"Please copy the procedures written on the chalkboard and we will go over them when you finish."

1. To subtract a fraction or mixed number from a whole number, take one (1) from the whole number and change it into a fraction making the numerator and common denominator the same. Then subtract fractions and subtract whole numbers.

$$
7 - 1\frac{7}{16} \;=\; 6\frac{16}{16} - 1\frac{7}{16} \;=\; 5\frac{9}{16}
$$

2. To subtract mixed numbers, first find the common denominator of the fractions, then subtract the fractions, and then subtract the whole numbers.

$$6\ \frac{4}{5}\ =\ 6\ \frac{4}{5}\ \times\ \frac{2}{2}\ =\ 6\ \frac{8}{10}\quad\text{(minuend)}$$

$$-\ 3\ \frac{1}{2}\ =\ 3\ \frac{1}{2}\ \times\ \frac{5}{5}\ =\ 3\ \frac{5}{10}\quad\text{(subtrahend)}$$

$$3\ \frac{3}{10}$$

3. If the fraction in the subtrahend is larger than the fraction in the minuend, take one (1) from the whole number in the minuend and increase the fraction.

$$6\ \frac{1}{8}\ =\ 5\ \frac{8}{8}\ +\ \frac{1}{8}\ =\ 5\ \frac{9}{8}$$

$$-\ \frac{7}{8}\qquad\qquad\qquad =\ \frac{7}{8}$$

$$5\ \frac{2}{8}\ =\ 5\ \frac{1}{4}$$

$$9\ \frac{1}{3}\ =\ 9\ \frac{2}{6}\ =\ 8\ \frac{8}{6}$$

$$-\ 4\ \frac{5}{6}\ =\ 4\ \frac{5}{6}\ =\ 4\ \frac{5}{6}$$

$$4\ \frac{3}{6}\ =\ 4\ \frac{1}{2}$$

"How did you get $5\ \frac{9}{8}$ from $6\ \frac{1}{8}$? Shouldn't it be $5\ \frac{8}{8}$?" asked Marsha.

$$6\ \frac{1}{8}\ =\ 5\ +\ \frac{8}{8}\ +\ \frac{1}{8}\ =\ 5\ \frac{9}{8}$$

"Don't forget you already had a fraction $\frac{1}{8}$ so you have to add $\frac{8}{8}$, which

you got from breaking the whole number 6 to 5 and $\frac{8}{8}$. You use the same

procedure with the second problem. You change the

$$\frac{2}{} \quad\text{to}\quad \frac{6}{} \ +\ \frac{2}{} \ =\ \frac{8}{} \quad\text{"}$$

9   6   8   6   6   8   6.

"Please work these examples with your partners."

a) $\frac{7}{8}$ - $\frac{3}{8}$
            6   2

b) $\frac{1}{3}$ - $\frac{3}{4}$
            5   2

"Lhyn, how did you do the first problem?"
"I don't like to write my problem out vertically, so I did it sideways. I changed the $6\frac{7}{8}$ to an improper fraction and I did the same with $2\frac{3}{8}$"

$$6\frac{7}{8} - 2\frac{3}{8} = \frac{55}{8} - \frac{19}{8} = \frac{36}{8} = 4\frac{4}{8} = 4\frac{1}{2}$$

"May I come up to the board and do the next problem? I worked with Lhyn so I did it the same way," said Sean.

$$5\frac{1}{3} - 2\frac{3}{4} = \frac{16}{3} - \frac{11}{4} = \frac{64}{12} - \frac{33}{12} = \frac{31}{12} = 2\frac{7}{12}$$

"Boy, that sideways business is confusing to me, so I did it the other way. I got the same answer!" said Darryl.

"Please do the problems, which ever way is easier for you to visualize and think—horizontally or vertically. You may now get into groups or work with a partner for practice on these fractions. We will have a test on Friday on subtraction of fractions. Let's have a small study group up front for those who need extra help."

"Will you please do your own work first, and then check with your partner to see if you have the same answers. If your answers differ, work out the problem together again and see if you derive a common answer."

Michael threw a ball of paper into the wastebasket from his seat. 'Oh, oh, that's a detention," the class declared.

"I'm sorry, I'm sorry! I'll pick it up," said Michael.

"Michael, I know what fun it can be to practice your shot put because basketball is my favorite sport, too. But you know that disturbing the class during study time is a 'No, no!'"

"Yes, Michael, it's a 'No, no!'" the class laughingly said.

"I'll go and pick all the papers from the floor," said Michael.

(Suddenly, several students started to drop papers on the floor. Michael picked them all up smiling without a complaint.)

"May I throw out the wastebasket into the big container by the office?" asked Michael.

(Michael has a difficult time staying in his seat for an extended time.

Instead of reprimanding him every time he gets up to walk, I have found giving him permission to walk in the room without disturbing any student has helped him. Now, he needs to go *out* of the classroom—to walk out in the empty hallway, away from people.)

"Yes, Michael. Thank you for being so helpful."

"I think we need some practice in regrouping. Will the paper monitors please pass out this worksheet. You are to write the missing numerators, class.

$$1 = \dfrac{-}{5} ; \quad 2 = 1\dfrac{1}{2} ; \quad 1\dfrac{1}{4} = \dfrac{-}{4} ; \quad 3\dfrac{3}{5} = 2\dfrac{-}{5} ;$$

$$7\dfrac{5}{8} = 6\dfrac{-}{8} ; \quad 1\dfrac{3}{4} = 1\dfrac{-}{8} ; \quad 4\dfrac{3}{4} = 3\dfrac{-}{12}.$$

Skill Practice (noting specific problems).

Examples of fractions and mixed number subtraction.

$\dfrac{4}{5} - \dfrac{1}{5}$ Subtraction of like denominators.

$\dfrac{11}{16} - \dfrac{5}{16}$ Subtraction of like denominators.
Reduction in answer.

$\dfrac{7}{8} - \dfrac{1}{4}$ Subtraction of unlike denominators.

$6\dfrac{4}{5}$
$-\ 3\dfrac{1}{2}$ Subtraction of mixed number.
Change fraction to equivalent fractions.

$$7\ \frac{2}{3}$$
$$-\ \ 3$$

Subtraction of whole number from mixed fraction.

$$8$$
$$-\ \ \ \frac{7}{16}$$

Subtraction of mixed number from whole number.

$$6\ \frac{1}{8}$$
$$-\ \ \ \frac{7}{8}$$

Fraction in the subtrahend larger than in minuend.

$$9\ \frac{1}{5}$$
$$-\ 4\ \frac{3}{6}$$

Mixed number subtraction with different denominators. Fraction in subtrahend larger than in minuend—equivalent fraction.

CHAPTER 11

SET UP 11: MULTIPLICATION OF FRACTIONS AND MIXED NUMBERS

This objective requires students to multiply fractions with fractions, whole number with fraction, mixed number with fractions, and mixed number with mixed number; utilization of cancellation procedure where applicable.

TEST PROBLEMS

a) $\dfrac{5}{6}$ x $\dfrac{1}{6}$

b) $\dfrac{3}{5}$ x $\dfrac{5}{8}$

c) 6 x $\dfrac{1}{3}$

d) $\dfrac{1}{2}$ x $1\dfrac{1}{2}$

e) $3\dfrac{3}{8}$ x $1\dfrac{7}{9}$

LIST OF COMMON ERRORS:

1. Computation errors.
2. Failure to change a whole number to a fraction.
3. Error in cancellation.
4. Inverted Multiplier.
5. Inverted Multiplicand.

6. Added numerators and multiplied denominators.

7. Error in reduction of proper and improper fractions.

8. Found common denominator.

EFFECTIVE STRATEGY—Mastering of Multiplication of Fractions with Fractions

(while carrying on dialogue with students).

"The process of multiplication of fractions is not difficult. I think most of you already know how to do these problems, so let's get started."

$$
\text{a)} \quad \frac{1}{4} \times \frac{1}{2} \qquad\qquad \text{b)} \quad \frac{1}{8} \times \frac{1}{3}
$$

$$
\text{c)} \quad \frac{5}{8} \times \frac{7}{5} \qquad\qquad \text{d)} \quad \frac{9}{10} \times \frac{3}{8}
$$

$$
\text{e)} \quad \frac{4}{5} \times \frac{3}{4} \qquad\qquad \text{f)} \quad \frac{5}{6} \times \frac{6}{7}
$$

"What did you get for problem 'a'?"

"$\frac{1}{8}$; $\frac{1}{4}$," responded different groups.

"Who got $\frac{1}{4}$?"

(Four hands went up.)

"Will you come up and show us how you got $\frac{1}{4}$, Marsha."

Tiffany said, "I'll do it. Marsha and I worked together. It was easy!"

"I'll come up to show you for me and Jon's work," said Bryan.

(Bryan came up to show his work with Jonathan. Tiffany followed him to the board to show her work.)

Tiffany said, "1 + 1 (pointing to the numerator) = 2; x 2 = 8. Then, I reduced the answer, $\frac{2}{8}$ to $\frac{1}{4}$ See, I remembered to reduce!"

"I can see how you made that mistake. We have been adding fractions until now, so you added the $\frac{1}{4}$ x $\frac{1}{4}$ You might circle the operation sign (x) in the

middle or write the problem like this: $\frac{1}{4}$ x $\frac{1}{2}$ = $\frac{1}{8}$."

"Will you please continue with the next problem, Bryan?"

"Okay, then it's $\frac{1}{24}$; $\frac{7}{8}$; $\frac{27}{80}$; $\frac{3}{5}$; $\frac{5}{7}$, " said Bryan.

"I got '0' for the last two problems," said Kenya.
"Come up and show us how you got '0', Kenya."

"$\frac{4}{5}$ x $\frac{3}{4}$ I cancelled the 4's, so I have '0' left. 3 x 0 = 0 and 5 x 0 = 0

said Kenya.

"I don't like to use the word 'cancel' because it may be interpreted as cancelled out, nothing left. What you are doing is finding the greatest common factor that the numerator and the denominator is 'divisible by' to simplify the computation. So, 4 divided by 4 is 1 and not 0. The numerator would then be 1 x 3 = 3 in the numerator; the denominator would then be 5 x 1 = 5. The answer would be $\frac{3}{5}$.

You can also multiply straight across, 4 x 3 = 12, the numerator; 5 x 4 = 20 the denominator. Then, you can simplify $\frac{12}{20}$ for your final answer by dividing the numerator and the denominator by 4, which results in $\frac{3}{5}$"

"We don't want to cross cancel. It only confuses us, so we're going to reduce the fraction at the end," said Bryan and Patrick."
"You may use whatever method is understandable or easier for you. Just remember to simplify the final answer because the achievement tests will have Let's go on. This time we will multiply whole number by fraction."

a) $\dfrac{5}{6}$ x 6   b) $\dfrac{5}{8}$ x 16

c) 12 x $\dfrac{5}{6}$ d) 30 x $\dfrac{2}{5}$

"Remember when you reduced improper fractions like $\dfrac{6}{1}$ = 6?

In the following problems, we are going to do just the opposite. We are going to change the whole number to a fraction, so the denominator will be 1."

$$5 = \dfrac{5}{1} \; ; \qquad 15 = \dfrac{15}{1} \; ; \qquad 150 = \dfrac{150}{1}$$

$$5 \times \dfrac{1}{5} = \dfrac{5}{1} \times \dfrac{1}{5} = \dfrac{5}{5} = 1; \qquad 15 \times \dfrac{3}{15} = \dfrac{15}{1} \times \dfrac{3}{15} = \dfrac{45}{15} = 3$$

"Wow, you all got these problems correct! We are really moving forward fast!"

"We're smart, right? And we are doing our homework, too!" said Mark.

EFFECTIVE STRATEGY—Mastering_of fractions and mixed numbers (while carrying on dialogue with students).

Procedure for multiplying fractions and mixed numbers:
   a) Change a whole number to the fractional form by using the whole number as the numerator and 1 as the denominator.

   b) Whenever it is possible, first divide any numerator and denominator by the greatest possible number that is exactly contained in both to simplify computation.

   c) Multiply the resulting numerators to obtain the numerator of the answer, and multiply the resulting denominators to obtain the denominator of the answer.

   d) Change each mixed number to an improper fraction.

   e) Reduce the answer to simplest form.

   Please copy this into your notes so you can refer back to it when you forget a step."

"This time we will multiply mixed fractions. Remember to change them to improper fractions first."

a) $1\frac{1}{4}$ x $1\frac{3}{5}$ b) $5\frac{1}{3}$ x $4\frac{1}{2}$

c) $3\frac{3}{4}$ x $1\frac{1}{5}$ d) $2\frac{1}{8}$ x $1\frac{1}{2}$

"Let's do the first problem together.

$$1\frac{1}{4} \times 1\frac{3}{5} = \frac{5}{4} \times \frac{8}{5} = \frac{40}{20} = \frac{8}{4}$$

"Why didn't you turn the $\frac{8}{5}$ upside down in problem a? Aren't you suppose to?" asked Keesha.

"That's in division. We'll get to that next."

"Please check the following answers with me:

$$5\frac{1}{3} \times 4\frac{1}{2}$$

$$\frac{\overset{8}{\cancel{16}}}{\underset{1}{\cancel{3}}} \times \frac{\overset{3}{\cancel{9}}}{\underset{1}{\cancel{2}}} = \frac{24}{1} = 24$$

Some of you might find it easier to write the improper fraction right below the mixed number so you can keep track."

"Mrs. Yamate, can you reduce two numerators by the same number, like

if you have $\frac{\overset{1}{3}}{5}$ x $\frac{\overset{3}{9}}{10}$ can I reduce the 3 to 1 and the 9 to 3?" said Patrick.

"That is a good question, Patrick. So many students get confused on can-cellation. No. Let's look over our procedure again. It says in 'b' first divide any numerator and denominator…the numerator and the denominator may be right on top or below each other or across each other."

$$\frac{3}{9} \quad \text{or} \quad \frac{3}{9} \times \underline{}$$

"Okay, I can't cancel out $\underline{3}$ x $\underline{9}$ to become $\underline{1}$ x $\underline{3}$ in the numerator
　　　　　　　　　　　5　10　　　　　1 x 2 in the denominator.

"You should have 4 $\underline{1}$ for problem 'c' and 3 $\underline{3}$ for the last."
　　　　　　　　　　2　　　　　　　　　　16

"You know, I've been thinking, why do I end up with a smaller number when
I multiply? It's suppose to get larger...like $\underline{1}$ x $\underline{1}$ = $\underline{1}$. That's smaller
　　　　　　　　　　　　　　　　　　2　3　6
than $\underline{1}$ and I multiplied it by $\underline{1}$ " said Charles.
　　　2　　　　　　　　　　　3

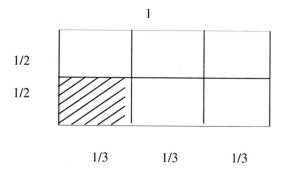

"1/3 of 1/2 is 1/6. The product is smaller than the given number because a
fractional part of the given number is being found."

"We haven't had a timed test for awhile and I think it might be fun to see
how well we understand multiplication of fraction before our test. How do you
feel about it, class?"

"Are you going to grade us?" said Jimmy.

"It's up to you. Why don't we decide after the mini test. If you do well on
it, you may want to count it in your average grade. See how many of these
problems you can complete in 5 minutes."

1. 2 1/3 x 21   4. 2 4 x 3 1/8   7. 4 1/4 x 2 1/8

2. 6 1/4 x 12   5. 6 0 x 3 1/12   8. 3 1/4 x 2 2/3

3. 4 1/5 x 15   6. 4 9 x 2 1/7   9. 18 2/3 x 1 3/4

Skill Practice (noting specific problems).

7/8 x 3/5	Multiply the numerators to get the numerator of the product and multiply the denominators to get the denominator of the product.
2/3 x 5	Whole number has to be changed to fractional form.
4/5 x 15/16	The numerator and the denominator can be simplified before computation done.
2 1/2 x 1 1/5	Mixed numbers need to be changed to improper fractions.

"Please pick up your homework worksheets. Anyone needing conferences time please sign up now."

"I'm glad to see that those of you who didn't finish the mini test in five minutes drew a line across your papers and finished it. Good discipline! Finish what you start!"

(Accountability. Respect for individual pacing.)

"I like doing multiplication of fractions. It's easy," said Marsha." I know I am going to get an 'A' on tomorrow's test."

CHAPTER 12

SET UP 12: DIVISION OF FRACTIONS AND MIXED NUMBERS

This objective requires student to divide fractions and mixed numbers.

TEST PROBLEMS

a) $\dfrac{3}{4}$ $\div$ $\dfrac{4}{8}$

b) 1 $\div$ $\dfrac{1}{2}$

c) 16 $\div$ $3\dfrac{1}{5}$

d) $4\dfrac{3}{8}$ $\div$ $\dfrac{5}{12}$

e) $2\dfrac{4}{5}$ $\div$ $\dfrac{7}{8}$

LIST OF COMMON ERRORS:

1. Wrong process—used multiplication.
2. Computation errors.
3. Inverted dividend.
4. Inverted dividend and divisor.
5. In complete cancellation got zero for the quotient.
6. Failure to reduce in answer.
7. Divided fraction by fraction and whole by whole.

EFFECTIVE STRATEGY—Mastering of Division of Fractions and Mixed Numbers).

"Class, do any of you remember what a multiplicative inverse or reciprocal is?"
"Never heard of it," responded the class.

The following was written on the blackboard:

6 and $\dfrac{1}{6}$ are multiplicative inverses of each other because

$\dfrac{6}{1}$ x $\dfrac{1}{6}$ = 1

$\dfrac{4}{3}$ and $\dfrac{3}{4}$ are multiplicative inverses of each other because

$\dfrac{4}{3}$ x $\dfrac{3}{4}$ = 1

Multiplying or dividing both the numerator and the denominator of a fraction by the same number does not change the value of the fraction.
(Blank faces. Are they understanding this?)

10 ÷ $\dfrac{2}{3}$ may be written $\dfrac{\frac{10}{1}}{\frac{2}{3}}$ If you multiply the

denominator $\dfrac{2}{3}$ by $\dfrac{3}{2}$, the reciprocal, the denominator will become 1.

$$\dfrac{\dfrac{10}{1} \text{ x } \dfrac{3}{2}}{\dfrac{2}{3} \text{ x } \dfrac{3}{2}} = 15$$

fraction   x reciprocal of fraction

The numerator must also be multiplied by 3/2 so as not to change the value of the fraction. The fraction 3/2 is used as the multiplier of both numerator and the denominator because it is the reciprocal of the divisor 2/3.

"Do we have to know all them stuff on the board, Mrs. Yamate? I don't understand it. It's confusing," said Sean.

"Yeah," said the class. "Do we have to learn this?"

"I get the last line, though," said Mark. I know how to divide fractions."

"I want you to understand that in order to divide a fraction by a fraction, you multiply the dividend by the reciprocal of the divisor. You'll understand it when we start to do the division of fraction."

"Please do the problems on the chalkboard."

a) $\dfrac{1}{2} \div \dfrac{2}{3}$ d) $\dfrac{7}{8} \div \dfrac{5}{12}$

b) $\dfrac{11}{12} \div \dfrac{4}{7}$ e) $\dfrac{7}{8} \div 7$

c) $\dfrac{2}{3} \div \dfrac{2}{5}$

"May we work together on this?" said Stephen.

"I would prefer that you do these alone so I can check any errors. Some of you may have some misconception. May I have the answer for 'a'?"

"2/6; 1/3; 3/4," replied different students.

"Sean, how did you get 2/6?" said Jamal.

"Andrew, you forgot to reduce to 1/3," said Lhyn.

"Wait a minute, Lhyn. Let's have Andrew explain his work."

"It's easy. I just multiplied 1/2 x 2/3. I remember my teacher telling us to change the division to multiplication," said Andrew.

"Andrew, you do multiply the dividend, but by the *reciprocal* of the divisor. Invert the divisor to find the reciprocal 2/3 x 3/2 = 1; therefore, 3/2 is the reciprocal," said Jamal.

"Which is the dividend and which is the divisor?" said Andrew.

$$\frac{1}{2} \quad \div \quad \frac{2}{3} \qquad\qquad \frac{1}{2} \quad \div \quad \frac{2}{3}$$

dividend   divisor

$$\frac{1}{2} \quad x \quad \frac{3}{2} \; = \; \frac{3}{4}$$

"Class, please rewrite the problem right underneath. I think it will keep your mind on the right track. When you rewrite the problem right on top, you cannot read the numbers clearly. I know, it takes a little time but the process will sink into your heads better."

"Now, Andrew, will you do the problem over and give us your answer."

"3/4," said Andrew.

"Did you get 1 29/48 for 'b'?"

"Yes." "No." "Forgot to reduce." Various answers were echoed.

"When you have an improper fraction like 77/48, you know there is 1 whole (48/48); write down the 1 and subtract 48 from 77 for the remainder. The answer is 1 29/48."

"The answer for 'c' is 1 2/3, and 'd' is 2 1/10," said Lhyn.

"Very good!"

"I got 1 3/5 for 'd'," said Keesha. 7/8 ÷ 5/12 = 84/40 and I reduced this one all the way! 1 24/40 = 1 12/20 = 1 6/10 = 1 3/5," said Keesha.

"Your reduction of 1 24/40 is correct, Keesha. Next time you might try taking the greatest common factor so you won't have to go through so many steps...3 x 8 = 24 and 5 x 8 = 40, so use 8."

"No, I would rather be sure...dividing by 2 is okay," said Keesha.

"Keesha, please take a step back to 7/8 x 12/5 = 84/40 = 2 4/40 = 2 1/10."

"Oh, it's 7 x 12 = 84 and 8 x 5 = 40. I know where I made the mistake," said Keesha. "I'll correct it. Wait."

"What do you have for 'e'?"

"8." "1/8." "6 1/8," different replies.

"The person who got '8' and '6 1/8' please come up to the board and show us how you solved your problem."

$$\frac{7}{8} \div 7 \qquad \frac{\overset{1}{8}}{\underset{1}{7}} \; x \; \frac{7}{1} \; = \; \frac{8}{1} \; = 8 \qquad\qquad \frac{7}{8} \; x \; \frac{7}{1} \; = \; \frac{49}{8} \; = 6 \; \frac{1}{8}$$

Stephen's work   John's work

"Well, I know my answer is wrong now after seeing how you did the other problems…but this is what I had on my paper," said Stephen, who wrote on the left side of the board.

"All right, Stephen, we'll help you. Let's do the problem over again."

$$\frac{7}{8} \div 7 = \frac{7}{8} \times \frac{1}{7} = \frac{7}{56} = \frac{1}{8}.$$

(He did it correctly and the class clapped for him.)

"John, you correctly changed the whole number into a fractional form like you did in multiplication. What did you forget to do with the second fraction?"

"Turn it upside down," said Eric, John's partner.

(They corrected their work. The class clapped for them, too.)

"Class, the reason you can invert the divisor to find the reciprocal is because division by a number gives the same result as multiplication by the reciprocal of that number."

$$\frac{7}{8} \div 7 \text{ can be written: } \frac{\frac{7}{8}}{7} = \frac{\frac{7}{8} \times \frac{1}{7}}{\frac{7}{1} \times \frac{1}{7}} = \frac{1}{8}$$

$$\frac{1}{4} \div \frac{1}{4} = \frac{1}{4} \times \frac{4}{1} = \frac{1}{1} = 1$$

"You should have '10' for the last answer."

"I didn't hear everyone's voice answering to the problems. Does this mean that some of you had the wrong answer? I think we had better have a group come up to the front and work on the chalkboard. Those of you who understand can work together on page 435 in the textbook. Do just the even numbers. We'll do the odds for homework."

"I'm going to write the procedure for operation for the division of fractions and mixed numbers on the blackboard, so please copy it."

Procedure for division of fractions and mixed numbers.

1. Change mixed number to an improper fraction.

2. Change whole number into fractional form.

3. Invert the divisor to find the reciprocal and then multiply as in the multiplication of fractions.

4. Reduce where possible in your answer and within your problem.

"Today, we are going to do division of mixed numbers. Change each mixed number to an improper fraction."

a)   5 7/8 ÷ 8/3   b)   2 1/2 ÷   1/2

c) 20 3/4 ÷ 2 3/40   d)   2 1/2 ÷   3/4

e) 2 3/4 ÷   2   f)   17 ÷   2 1/3

"Shall we do the first problem together?"
"No, we can do it by ourselves," replied the class.
"What did you get for 'a'?"
"2 13/64," said Charles.
"I didn't get that," said Eric.
Charles came to the chalkboard and showed his work:
5 7/8 ÷ 8/3
47/8 x 3/8 = 141/64 = 2 13/64."
"I made a mistake in multiplication," said Tyshenna.
"The answer for 'b' is 5."
"Correct."
"I'll write the answer for 'c': 20 3/4 ÷ 2 3/40 = 83/4 x 40/83 = 10/1 =10 "Mrs. Yamate, you told us to write the multiplication under the division so we won't get mixed up. You didn't," said the class.
"Shucks, you caught me. Another mistake by the teacher! Sorry," (I rewrote the problem. Teacher can make mistakes, too!)
Andrew got up and started moving about. "Andrew, are you swimming or dancing?"
"Sorry, Mrs. Yamate. I'm tired of sitting in my seat," said Andrew.
"Why don't you all get up and stretch out. We have been doing a lot of concentrated studying. No wonder Andrew is restless."
"Let's finish the board work. What did you get for 'd' and 'e'?
" 3 1/3 for 'd' and 1 3/8 for 'e,' responded the class. "And we have 7 2/7 for 'f.' Right?"
"Correct me if I am mistaken, but I think all of you know how to do the division of fractions and mixed numbers now. Congratulations!"

"The reason why we made mistakes is because like me, I forgot which numbers to turn upside down. All the teachers told us was to memorize. Now, you are telling us why you invert," said Samuel, "and that's good."

(The key to their continuing mistakes… learned by rote.)

"Ready for a timed quiz in division of mixed numbers and fractions?"

"Yes."

"You will be given 5 minutes to do these problems:

1. $12 \div 2\ 1/3$	5. $24 \div 1\ 1/8$	9. $4\ 2/3 \div 1\ 1/2$
2. $20 \div 2\ 1/4$	6. $60 \div 2\ 1/4$	10. $2\ 1/3 \div 2\ 1/7$
3. $4\ 1/3 \div 5$	7. $49 \div 2\ 1/3$	11. $4\ 1/4 \div 2\ 1/8$
4. $6\ 1/4 \div 2$	8. $100 \div 2\ 2/3$	12. $3\ 2/3 \div 1\ 5/6$

"When 5 minutes is up I want you to put a line across your paper and finish it up like we did before. Some of us need a little more time to finish our work, and that is all right. I am sure you will gain speed with more practice."

Antarra came over to do his work at the end of my desk. "What's the matter? Do you need help?"

"I just want to sit here and do my work incase I forget something," said Andrew.

Jamal and Charles finished the quiz and came up to ask if they may correct the quiz for the class. "Correct our papers for us, Mrs. Yamate. Then me and Charles can help you correct the rest of the papers. We'll be your Teaching Assistants again."

"Thank you, Jamal and Charles."

Lhyn and Jennifer finished the quiz. "Hey, it isn't fair that you guys get to be Teaching Assistants again. It's the girl's turn," said Jennifer.

"Okay. Mrs. Yamate do you have two or more red pencils? We can all help," said Jamal.

"Yes, thank you. With the four of you coaching the class, we should be able to go on to the next topic tomorrow. You are getting closer to the algebra book!"

"Hey you guys, raise your hand if you need help. Me, Charles, Lhyn, and Jennifer will come around and help you. You know where we are heading!" said Jamal.

"Class, please study and do your worksheets on division of mixed numbers tonight. We are ready for our test for Friday, aren't we? Please sign up for conference if you need additional help."

"We have been studying division of fractions and mixed numbers. Have any of you noticed that when a number is divided by a fraction, the quotient is always greater than the given number?" Example: 3 ÷ 3/4
$$3 \times 4/3 = 4$$

"Yeah, why is it?" asked the class.

"When a number is divided by a mixed number, the quotient is always smaller than the given number." Example: 3 1/4 ÷ 1 5/8
$$13/4 \times 8/13 = 2/1 = 2$$

Skill Practice (noting specific problems)

$\frac{2}{3} \div \frac{5}{6}$		a) Invert divisor and simplify.
$8 \frac{3}{4} \div 7$		b) Change mixed number to improper fraction, change whole number to fraction form, and simplify. Invert divisor.
$1\,5 \div \frac{3}{8}$		c) Change whole number to fractional form and simplify. Invert divisor.
$2 \frac{1}{2} \div \frac{3}{4}$		d) Change mixed number to proper fraction. Invert divisor. Simplify.
$2 \frac{3}{16} \div 1 \frac{1}{4}$		e) Change mixed numbers to improper fractions. Invert divisor. Simplify.

"I see that you are all helping each other with today's assignment before I told you. You have all become helpful and concerned for each other's learning. I am proud of you all! When I first received the class list for the "remedial mathematics class," I was worried that I may not be able to reach you and get you to learn. Remember the way you looked at me when you entered my class the first day of school?"

(Students were looking at each other and smiled...guilty?)

"Today is Thursday, so let's have conference now. You are ready for a test on division of fractions, so may we take it tomorrow?"

CHAPTER 13

SET UP 13: ADDITION, SUBTRACTION, MULTIPLICATION, AND DIVISION USING EXPONENTS

This objective requires students to add, subtract, multiply, and divide a number with exponent.

TEST PROBLEMS

a) 3^2 + 5^2

b) 3^2 -   5

c) 2^2 x   5

d) 100 ÷ 5^2

LIST OF COMMON ERROR:

Multiplied base times exponent.

EFFECTIVE STRATEGY—Mastering of Exponents (while carrying on dialogue with students).

a) $3^2 + 6^2 + 5^2$ e) 3^2 x 5

b) $4^3 + 2^4 + 3^5$ f) 9^2 x 3

c) $10^2 + 10^3$ g) 192 ÷ 8^2

d) $4^2 - 9$ h) 54 ÷ 3^3

"Please do the problems on the board. Do you remember these problems in your achievement test?"

"Yes. But we never saw those little numbers on the top before so we didn't do them. Hey, did any of you know what that stuff was?" asked Anthony.

"Nope. I didn't do them," said Gregg.

"What did you get for the first problem?"

"2 8," responded the class (All in agreement).

"How did you get 2 8?"

"Easy. 3 x 2 = 6; 6 x 2 = 12; 5 x 2 =10. It adds up to 28," said Matt.

"All right. Let's look at 3^2 . The number 3 (factor) is called the *base*. The small numeral written to the upper right is called an *exponent*. It tells how many times the factor is being used. When two equal factors are used in multiplication like 3 x 3, it may be written in exponential form as 3 to the second power. It is read 'three to the second power' or 'three squared', 'the second power of 3', or 'the square of 3'." (Talking and writing on the chalkboard)

"What is 3^2 ?"

"9," responded the students.

"What is 6^2?"

"6 x 6 = 3 6," responded the students.

"What is 5^2 ?"

"5 x 5 = 2 5," responded the students.

"Now add the numbers."

"7 0," responded the students.

"I thought that they made a mistake on the test when I saw that small number. I've never seen it before the test," said Jon.

"What do you think 4^3 means?"

"4 x 3?" asked Douglas.

"No. 4 x 4 x 4," said Fred. "I get it now."

"4 x 4 x 4 = 6 4. It is read 'four to the third power', 'four cubed', or 'the cube of four'." (Talking and writing on the chalkboard)

"What did you get for 2^4?"

"2 x 2 x 2 x 2 = 1 6. Right?" asked Matt and Stephen.

"What did you get for 3^5 ?"

"3 x 3 x 3 x 3 x 3 = 2 4 3," said Walter.

"You can multiply 9 x 9 x 3. I think it'll be faster," said Marsha and Latania.

"Good thinking! Now multiply the factors and add the products."

"323," responded Jennifer and Lhyn.

"What did you get for 'c'?

"1100. Hey, if you have a 10 and an exponent of 2, you will have two zeros, $10^2 = 1\ 0\ 0$; and if you have a 10 and an exponent of 3, you will have three zeros, $10^3 = 1\ 0\ 0\ 0$," said Naomah and Deidre.

"Oh yeah, you guys are right. This is easy!" said Marvin and Michael.

"Numbers like 10^2, 10^3, 10^4…are called powers of 10."

"We'll give out the answers," said Jamal and Charles. "Everyone check your answers now. Is that okay with you, Mrs. Yamate? We know this now."

"Okay, boys. Take over."

"If you guys are going to read off the answers, you better let the girls read the answers next time," said Lhyn.

"Hey, let me read the answers," said Michael.

"Okay, let Michael read. He's been doing real good now," said Jamal.

"Did you all get 7 for 'd'?" asked Michael, standing confidently in front of the class.

"Yes," responded the class.

"Did you get 4 5 for 'e'? 2 4 3 for 'f'? 1 2 for 'g'? and 2 for 'h'? asked Michael. "Put an 'A' on top of your paper if you got them all correct. I'll come and initial your grade."

(Interesting that he should want to initial their classmates' grade…and the classmates are accepting this. Michael has gained self-confidence!)

The following was written on the blackboard:

Find the value of the following:　　a) 16^2　　　c) 2^8

　　　　　　　　　　　　　　　　　b) 4^4　　　　d) 3^7

Use the exponential form to write: a) $2 \times 2 \times 2 \times 2$　　　c) $10 \times 10 \times 10$

　　　　　　　　　　　　　　　　b) $8 \times 8 \times 8$　　　　d) $6 \times 6 \times 6 \times 6 \times 6$

Write as a numeral:　a) Two the eighth power　　c) Nine cubed

　　　　　　　　　　b) Fifty squared　　　　　　d) Five to the sixth power

"We're going to write the answers on the chalkboard now. Ready? said Lhyn and Jennifer."

"We don't want you to get stressed out, Mrs. Yamate, so we'll take over from here," said Douglas and Fred. "You can sit down now, so you won't get arthritis."

"Where did you hear about arthritis?"

"Oh, my grandmother has arthritis so she has a hard time walking," said Douglas. "We don't want your hands to get arthritis."

"All right. Thank you Doug and Fred."

(Their concern for my health is heart-warming!)

"This was easy. It looked so hard at first," said Keesha.

"You may start your homework worksheet now. If you have any questions, please see your leaders. I am going to take a rest! I think we can take a test on Thursday, right?"

"Mrs. Yamate, will you teach us stuff out of the red book on your desk? (Introduction to Algebra) Aren't we through with the review?" said Charles.

"Yeah, we want to learn algebra like those smart kids," said Jamal.

"We can skip the Pre-Algebra book," said Samuel.

"Will you give me time to do some re-assessment—of what you have learned thus far? I need to have you take an achievement test similar to what you took at the beginning of my class. Would you agree to that?"

"All right. We trust you, Mrs. Yamate," responded Jamal. "Right, class?"

"You have proven to me that you can take on the challenge. So, as soon as we finish the achievement test, we will begin to study out of the Algebra book. You see, you will need to use the principles and rules you learned in arithmetic for doing problems in Algebra."

"Okay, you guys. Let's all study and help each other so we can get into the red book. Mrs. Yamate said she will teach us," said Charles.

"That will be your reward for being so conscientious about doing your review work and being so cooperative."

"Okay, let's get into our study groups," said Charles and Jamal.

(The students are moving their desks and discussing how to go about studying in their groups.)

BIBLIOGRAPHY

Adams, Sam, L. D. Ellis, and B. F. Beeson. <u>Teaching Mathematics: With Emphasis on the Diagnostic Approach</u>. New York: Harper and Row Publishers, 1977.

Ashlock, Robert B. <u>Error Patterns in Computation.</u> Columbus, Ohio: Charles E. Merrill Publishing Co., 1976.

Bell, Frederick H. <u>Teaching Elementary School Mathematics: Methods and Content for Grades K-12</u>. Dubuque, Iowa: Wm. C. Brown Co., Publishers, 1980.

Bruekner, Leo J and Foster E. Grossnickle. <u>Making Arithmetic Meaningful</u>. Philadelphia: The John C. Winston Co., 1953

Copeland, Richard W. <u>How Children Learn Mathematics: Teaching Implications of Piaget's Research</u>. New York: McMillan Publishing Co., Inc., 1974.

_____. <u>Mathematics and the Elementary Teacher</u>. Philadelphia: W. B. Saunders Co., 1972.

Flanders, James R. "How Much of the Content in Mathematics Textbooks Is New?" <u>Arithmetic Teacher</u>, XXXV (September, 1987), 18-23.

Freeman, Donald J., G. M. Belli, A. C. Porter, R. E. Floden, W. H. Schmidt, J. R. Schwille. "The Influence of Different Styles of Textbook Use on Instructional Validity of Standardized Tests." <u>Journal of Educational Measurements</u>, 20 (Fall, 1993), 259-270.

Fulkerson, Elbert. "Adding by Tens", <u>Arithmetic Teacher</u>, XX (March, 1963).

George, Pamela. "Coaching for Standardized Tests: Efficacy and Ethics" <u>Mathematics Teacher</u>, XXC (September, 1987), 424-426.

Horvath, Patricia J. "A Look at the Second International Mathematics Study Results in the U.S.A. and Japan." <u>Mathematics Teacher</u>, XXC (May, 1987), 359-368.

McKnight, Curtis C. And Kenneth J. Travers. "Eighth-Grade Mathematics in U.S. Schools: A Report from the Second International Mathematics Study." rithmetic Teacher, XXXII (April, 1985), 20-26.

_____. "Twelfth Grade Mathematics in U.S. High Schools: A Report from the Second International Mathematics Study." Mathematics Teacher, LXXVIII (April, 1985), 292-300.

Rosner, J. Helping Children Overcome Learning Difficulties. New York: Walker and Co., 1975.

Schulz, Richard. "Characteristics and Needs of the Slow Learner," National Council of Teachers of Mathematics. Thirty-fifth Yearbook, 1972.

Travers, Robert M. An Introduction to Educational Research. New York: The McMillan Co., 1969.

Usiskin, Zalman. "Why Elementary Algebra Can, Should, and Must Be an Eighth Grade Course for Average Students." Mathematics Teacher, LXXX, (September, 1980), 428-438.

Wilson, G. M. "What Arithmetic Shall We Teach?" Boston Mass.: Houghton-Mifflin Co., 1926.

978-0-595-27736-0
0-595-27736-5

Printed in the United States
113543LV00006B/463-480/A